ELECTRICAL SYSTEMS DESIGN
DATA HANDBOOK

SECOND EDITION

ELECTRICAL SYSTEMS DESIGN

DATA HANDBOOK

Prof. M.K. Giridharan

Former Dean (Academics) & Professor
Department of Electrical & Electronics Engineering
Mar Baselios College of Engineering & Technology
Thiruvananthapuram, Kerala

APPROVED DATA HANDBOOK FOR UNIVERSITY EXAMINATIONS

TECHSAR™
NEW DELHI

Published and Printed by
TechSar Pvt. Ltd.
4435–36/7, Ansari Road, Daryaganj
New Delhi-110002 (India)
E-mail: tech_info@techsarworld.com
Website: www.techsarworld.com

ISBN 978-93-90620-78-4

Reprint 2025

This handbook is dedicated to my

Students

PREFACE TO THE SECOND EDITION

As a part of Intended Nationally Determined Contributions (INDCs), India has committed to increase the share of installed capacity of electric power from non-fossil fuel sources to 40% by 2030. Solar energy is one of the main sources to accomplish this target. Government of India has set the target of achieving 100 GW of solar power capacity in the country by the year 2022 of which 40 GW to be achieved from rooftop solar (RTS) plants.

The rooftop solar (RTS) plant is a system installed mainly on the roof of a building and includes installations on open contiguous land within the area of premises wherein valid and live electricity connection has been provided by the distribution utilities/companies (DISCOMS). Typically, 1 (one) kWp RTS plant requires about 10 m^2 area. The solar power so generated can then be used either for captive consumption of the premises or can be fed into the grid and adjusted in the electricity bill. Net metering regulations notified by respective State Electricity Regulatory Commissions (SERCs) provide a legal framework for such adjustment. RTS plants help DISCOMs in reducing transmission and distribution losses as power consumption and generation are co-located. These plants are also useful in tackling daytime peak load as solar generation profile matches such peak loads during the day.

The second edition of this well accepted reference handbook has two more sections added to its contents. They are the data regarding Solar PV systems for domestic applications and Emergency and Standby generation systems. With this addition, this data handbook will be able to provide more information to the budding electrical engineers, in their pursuit of better and more efficient systems design.

M K GIRIDHARAN

PREFACE TO THE FIRST EDITION

With the emerging trend of energy quality and efficiency, corporate facilities are demanding-greater quality and efficiency in Electrical Systems. This requires that the Electrical Systems designer to be more versatile and effective managers of energy resources. This data handbook is intended to be used as design assistance to the beginners in the field of Electrical Systems design and provides them an easy access to the relevant data required for their design.

This design data handbook is not intended for specialists in the field, but rather for the students of Electrical Engineering who are just entering the field of electrical systems design. This hand book also does not show the student how to be a designer, but present in a concise manner the basic reference data to perform the design functions.

This handbook will be of value to the engineering college students in electrical engineering,as it provides them with a good exposure of what is involved in electrical systems design and how to approach the design problem. This handbook can be permitted to be used inside the examination hall as a reference handbook.

Using this handbook will enable the students to perform many preliminary design studies and calculations that are necessary for designing. Computer software may allow the designer to acquire accurate answers to most of the design problems that would have been difficult or rather impossible with hand calculations. But for beginner to understand the basics it is always preferable to start from the basics with hand computations.

This handbook is to be used as a supplement to the book titled "Electrical Systems Design"by the same author. In conclusion, it must be stated that the full credit of this handbook is not due to the author, but to the various Standards Organizations and Companies who originated and developed those data. A humble attempt only is made to consolidate the data available else where, as it would be beneficial to the engineering students.

M K GIRIDHARAN

ACKNOWLEDGEMENTS

In the course of compiling the data for inclusion in this data handbook, I have selected data and information from several sources. This is done with a view to providing the new generation of engineering students, a comprehensive and easy to access source of design data in a condensed manner. Credits have been given to the sources, when information is derived from company websites, product catalogues or standard specifications. For this I am indebted to the following organizations and companies. The list is not complete.

Bureau of Indian Standards (BIS)
International Electrotechnical Commission (IEC)
Illumination Engineering Society (IES)
Institution of Electrical & Electronics Engineers (IEEE)
M/s Crompton Greaves India Ltd.
M/s Kirloskar Electric Company
M/s Finolex Cables Ltd.
M/s Havells India Ltd.
M/s Schnieder Electric India Pvt. Ltd.
M/s ABB Ltd.
M/s KSEB Ltd.
M/s Kerala State Electrical Inspectorate
M/s Leonics Company Ltd., Thailand
M/s Indsolar Pvt. Ltd.
M/s Exide Batteries

I would also like to put on record my heartfelt thanks to my ex-colleagues Ms Jomole Joseph and Ms Jeneena Babu of Mar Baselios College of Engineering and Technology, for their valuable assistance in correcting the drafts and suggesting modifications to the contents.

M K GIRIDHARAN

CONTENTS

TRANSFORMER SUBSTATIONS

CONDUCTORS AND CABLES

EMERGENCY AND STANDBY GENERATION

Table 1: S I UNITS AND QUANTITIES

Quantity	Quantity symbol	Unit symbol
Velocity	v	m/s
Mass	m	kg
Density	ρ	kg/m^3
Force	F	N
Work	W	J
Energy	E	J
Power	P	W
Efficiency	η	1
Electric charge	Q	C
Electrical potential	U	V
Electrical current strength	I	A
Electrical resistance	R	Ω
Length	l	m
Height	h	m
Radius	r	m
Diameter	d	m
Area	A	m^2
Time	t	s
Time constant	τ	s
Electric conductance	G	S
Specific resistance	ρ	Ω.m
Electric field strength	E	V/m
Electric current density	J	A/m^2
Electric capacitance	C	F
Dielectric constant	ε	F/m
Electric field constant	ε_0	F/m
Dielectric coefficient	ε_r	1
Magnetic flux	φ	Wb
Magnetic flux density	B	T
Magnetomotive force	Θ	A
Magnetic field strength	H	A/m
Magnetic field constant	μ_0	H/m
Permeability	μ	H/m
Permeability coefficient	μ_r	1
Number of turns	N	1
Thermodynamic temperature	T	K
Thermodynamic coefficient	α	1/K
Heat	Q	J
Specific heat	c	J/kg.K

Table 2: ALPHANUMERIC NOTATION, GRAPHICAL SYMBOLS AND COLOURS (National Electrical Code 2011)

Designation of Conductor	Alphanumeric notation	Graphical symbol	Colour
Supply system AC			
Phase 1	**L1**		Red
Phase 2	**L2**		Yellow
Phase 3	**L3**		Blue
Neutral	**N**		Black
Apparatus AC			
Phase 1	**U**		Red
Phase 2	**V**		Yellow
Phase 3	**W**		Blue
Neutral	**N**		Black
Supply system DC			
Positive	**L+**	+	Red
Negative	**L–**	–	Blue
Mid Conductor	**M**		Black
Supply system 1Φ			
Phase	**L**		Red
Neutral	**N**		Black
Protective earth	**PE**		Green & Yellow
Earth	**E**		No Colour
Clean earth	**TE**		
Chassis or frame	**MM**		
Equipotential terminal	**CC**		

Table 3: LIGHTING DESIGN PROCEDURES

The best option available for lighting design is the use of standard lighting software. If no such facility is available, any one of the following techniques can be used for the design.

- Short hand method
- Long hand method
- Simplified method

Short hand method

In this method it is assumed that the average maintained quantity of lumens arriving on the work plane will be half the quantity of the total new lamp lumens. This method assumes "normal" sized rooms, "normal" surface reflectance and "normal" dirt conditions. A normal sized room is the one in which the mounting height (distance from the luminaire bottom to the work plane) is less than half the smallest room dimension. It is also assumed that conventional fixtures are used. A high bay fixture is used when the mounting height is over 6 m. Otherwise low bay fixtures are recommended. The total number of fixtures (luminaires) to be used is then calculated as follows.

Step-1 Obtain the required level of illumination in lux.

Step-2 Choose a particular lamp and obtain the lamp lumens from the lamp data.

Step-3 Delivered maintained lamp lumen (DMLL) $= 0.5 \times$ Rated lamp lumens.

Step-4 Number of fixtures (Luminaires) $= \dfrac{\text{Room area} \times \text{Lux}}{\text{DMLL}}$

Select the nearest round figure.

Step-5 Area per fixture $= \dfrac{\text{Room area}}{\text{Number of fixtures}}$

Step-6 Spacing between fixtures $= \sqrt{\text{Area per fixture}}$

Spacing should not be more than the mounting height (MH) for high bay fixtures and 1.5 times the mounting height for low bay fixtures. Check for conformity. If the spacing is too large, go back to step-2 and start again with a lower lamp rating.

Long hand method

This is a more accurate and effective method of estimating the number of light fixtures (luminaires) and their spacing, because it takes into account the differences in Photometric performance caused by room geometry and system lumen depreciation. The core idea of this method is the determination of Coefficient of Utilization (CU) on the concept that the area to be lighted has three cavities or spaces that have effective reflectances with respect to each other and the work plane. The given space to be lighted is divided into three cavities that are shown below. The space between the ceiling and the bottom of the luminaire is known as Ceiling Cavity (CC) and the distance is h_{cc}, or the height of ceiling cavity. The space between the work plane and the bottom of luminaire is Room Cavity (RC) and the distance is h_{rc} or the height of room cavity. The space between the work plane and the floor is Floor Cavity (FC) and the height is h_{fc} or the height of floor cavity. The room cavities are shown in Figure 3-1.

The cavity ratios represent the geometric properties of the ceiling, room and floor cavities and can be found using the formula:

$$\text{Cavity ratio} = \frac{2.5 \times h \times (perimeter)}{Area}$$

where,

$h = h_{cc}$, for Ceiling Cavity Ratio (CCR)
$= h_{rc}$, for Room Cavity Ratio (RCR)
$= h_{fc}$, for Floor Cavity Ratio (FCR)

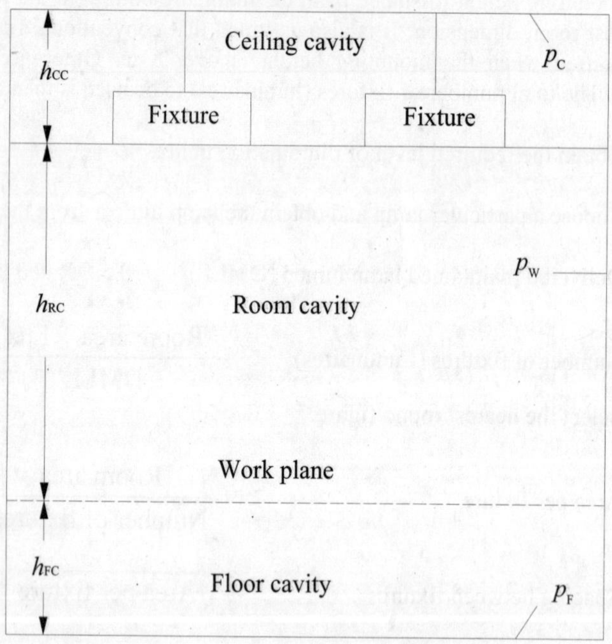

Figure 3-1 Details of room cavities

Step-1 Study the site plan and elevation of the installation. Determine the lighting needs in consultation with standards and the end user.

Step-2 Select the light sources and luminaire, appropriate to the installation geometry, nature of visual task and energy criteria.

Step-3 Calculate the ceiling, room and floor cavity ratios and using **Table 17** find out the effective ceiling and floor cavity reflectances.

Step-4 Use the given chart for coefficient of utilization, the effective ceiling and floor reflectances and RCR; find out the value of effective coefficient of utilization.

Step-5 Select the light loss factor (LLF), this factor was previously referred to as maintenance factor (MF). Generally, accepted values are:

A/C rooms, clean rooms, etc.	0.8
Industrial environment	0.7
Dusty areas	0.6

Step-6 Use the formula to arrive at the quantity of luminaires (fixtures)

$$\text{Number of Luminaires} = \frac{\text{Area} \times \text{Illumination}}{\text{CU} \times \text{LLF} \times \text{Lumen output of luminaire}}$$

Step-7 Arrange the luminaires symmetrically giving due considerations to spacing criterion to achieve uniformity of illumination.

$$\text{Area per luminaire (fixture)} = \frac{\text{Room area}}{\text{Number of fixtures}}$$

$$\text{Therefore, fixture spacing} = \sqrt{\text{Area per fixture}}$$

Spacing should not be more than the mounting height (MH) for high bay fixtures and 1.5 times the mounting height for low bay fixtures. Check for conformity. If the spacing is too large, go back to step-2 and start again with a lower lamp rating.

Simplified design procedure

A simplified design procedure for lighting design by lumen method for a simple interior is given below (Assumption: C.O.U. chart based on room index is available)

Step-1 Study the site plan and elevation of the installation. Determine the lighting needs in consultation with standards and the end user.

Step-2 Select the light sources and luminaire, appropriate to the installation geometry, nature of visual task and energy criteria.

Step-3 Calculate the room index (RI) using the formula:

$$\text{Room Index (RI)} = \frac{\text{Length} \times \text{Breadth}}{(\text{Length} + \text{Breadth}) \times \text{Mounting height}}$$

Step-4 Based on this, coefficient of utilization value can be obtained from the C.O.U. table for the selected luminaire. These values are generally based on the RI and room surface reflectances which are generally 70% for light coloured, 50% for average and 30% for dark walls and 50%, 30% and 10% for corresponding ceilings.

Step-5 Select the light loss factor (LLF), this factor was previously referred to as maintenance factor (MF). Generally, accepted values are:

A/C rooms, clean rooms, etc.	0.8
Industrial environment	0.7
Dusty areas	0.6

Step-6 Use the formula to arrive at the quantity of luminaires:

$$\text{Number of Luminaires} = \frac{\text{Area} \times \text{Illumination}}{\text{CU} \times \text{LLF} \times \text{Lumen output of luminaire}}$$

Step-7 Arrange the luminaires symmetrically giving due considerations to spacing criterion to achieve uniformity of illumination.

$$\text{Area per luminaire (fixture)} = \frac{\text{Room area}}{\text{Number of fixtures}}$$

$$\text{Therefore, fixture spacing} = \sqrt{\text{Area per fixture}}$$

If this spacing is too wide, repeat the process with a fixture with lower wattage of lamps. Normally, spacing should not be more than the mounting height for high bay fixtures and 1.5 times the mounting height for low bay fixtures.

Arrangement of fixtures

Fixtures shall be arranged from the middle of the room to the outside. A square array is the best option but rectangular array also will work satisfactorily as long as the spacing to mounting height ratio is not violated. With an odd quantity of fixture rows or columns, there will be a fixture line on the centre line. With even quantities of rows or columns, the locations are half the spacing off the centre line. The closest fixture to a wall should be one half the spacing or less.

Table 4: RECOMMENDED LEVELS OF ILLUMINATION FOR DOMESTIC BUILDINGS

(National Electrical Code 2011)

No.	Location		Illumination level-lux
1.	Entrances, Hallways		100
2.	Living room		300
3.	Dining room		150
4.	Bedroom	General	300
		Dressing, bed heads	200
5.	Kitchen		200
6.	Kitchen sink		300
7.	Bathroom		100
8.	Sewing		700
9.	Workshop		200
10.	Stairs		100
11.	Garage		70
12.	Study		300

Table 5: RECOMMENDED LEVELS OF ILLUMINATION FOR RECREATIONAL/ASSEMBLY BUILDINGS

(National Electrical Code 2011)

No.	Location	Illumination level-lux
1.	**Assembly and Concert Halls**	
	Foyers, auditoria	100-150
	Platforms	450
	Corridors	70
	Stairs	100
2.	**Cinema Halls**	
	Foyers	150
	Auditoria	50
	Corridors	70
	Stairs	100
3.	**Theatres**	
	Foyers	150
	Auditoria	70
	Corridors	70
	Stairs	100

Table 6: RECOMMENDED LEVELS OF ILLUMINATION FOR SCHOOL/COLLEGE BUILDINGS

(National Electrical Code 2011)

No.	Location	Illumination level-lux
1.	**Assembly halls**	
	General	150
	When used for examinations	300
	Platforms	300
2.	**Classes and lecture rooms**	
	Desktop	300
	Blackboard	20-300
3.	**Libraries**	
	Shelves	70-150
	Reading rooms	150-300
	Reading tables	300-700
	Cataloguing/stack rooms, etc.	150-300
4.	**General**	
	Office	300
	Staff rooms	150
	Corridors	70
	Stairs	100

Table 7: RECOMMENDED LEVELS OF ILLUMINATION FOR OFFICE BUILDINGS

(National Electrical Code 2011)

No.	Location	Illumination level-lux
1.	Entrance hall/reception	150
2.	Conference rooms/executive offices	300
3.	General office space	300
4.	Business machine operation	450
5.	Drawing office	450
6.	Corridors	70
7.	Stairs	100
8.	Lift landings	150

Table 8: RECOMMENDED LEVELS OF ILLUMINATION FOR HOSPITAL BUILDINGS

(National Electrical Code 2011)

No.	Location	Illumination level-lux
1.	Reception & waiting	150
	General wards	100
	Bed side	150
	Toilets	70
	Stairs	100
2.	**Operation theatre**	
	General	300
	Operating tables	special
3.	**Laboratories**	300
	Radiology	100
	Causality & OP	150
	Dispensaries	300
4.	**Laundry**	200
	Dry cleaning	200
	Ironing, mending	300
5.	General office	450
6.	Kitchen	200

TYPICAL LAMP DATA

Table 9: INCANDESCENT LAMPS

Rating–Watts	Life-hours	Initial Lumens
60	1000	870
100	750	1750
150-PAR	2000	1740
200	2000	2300
500	2000	6500

Table 10: FLUORESCENT LAMPS

Rating–Watts	Life-hours	Initial Lumens
18	7000	1120
20	7000	1020
36	7000	2800
40	7000	2700
2×40	7000	4000

Table 11: COMPACT FLUORESCENT LAMPS

Rating–Watts	Life-hours	Initial Lumens
5	10000	220
7	10000	380
11	10000	560
13	10000	680
15	10000	810
18	10000	1050
23	10000	1500
26	10000	1800
32	10000	2400

Table 12: LED LAMP

Company	Model	Watt	Lumens	L/W	PF	CRI	Life in hours	Warranty
Philips	Ace saver	2.5	230	92	0.25	80	15000	1 Year
	Ace saver	4	350	87.5	0.72	80	15000	1 Year
	Ace saver	7	600	85	0.85	80	15000	1 Year
	Ace saver	9	806	89	0.89	80	15000	1 Year
	Stellar Bright	12.5	1055	84	0.9	80	15000	1 Year
	Stellar Bright	14	1400	100	0.93	80	15000	1 Year
Havells	Adore	3	270	90	–	80	25000	1 Year
	Adore	5	400	80	–	70	25000	1 Year
	Adore	7	520	74	–	70	25000	1 Year
	Adore	10	800	80	–	70	25000	1 Year
	Adore	15	1500	100	–	80	25000	1 Year
Osram	Superstar Classic A	6	470	78	0.95	80	25000	1 Year
	Superstar Classic A	10	806	80.6	0.95	80	25000	1 Year
	Lightify Classic A	9.5	810	85	0.7	>80	25000	1 Year
	Superstar Classic A	11	1055	95	0.95	>80	25000	1 Year

Eveready	LED Bulb	3	300	100	0.5	80	20000	1 Year
	LED Bulb	5	500	100	0.5	80	20000	1 Year
	LED Bulb	7	700	100	0.5	80	20000	1 Year
	LED Bulb	9	900	100	0.9	80	20000	1 Year
	LED Bulb	12	1200	100	0.9	80	20000	1 Year
	LED Bulb	14	1400	100	0.9	80	20000	1 Year
Wipro	Garnet N50001	5	450	90	0.7	80	25000	1 Year
	Garnet N70002	7	610	87	0.8	80	25000	1 Year
	Garnet N90001	9	830	92	0.95	80	25000	1 Year
Oreva	ECO Series	4	316	79	—	—	15000	2 Years
	ECO Series	6	477	79.5	—	—	15000	2 Years
	ECO Series	8	615	76	—	—	15000	2 Years
	ECO Series	10	775	77.5	—	—	15000	2 Years
	ECO Series	12	935	77	—	—	15000	2 Years
	DX Series	11	925	84	—	—	15000	2 Years
Bajaj	LEDZ	5	400	80	0.7	—	25000	1 Year
	LEDZ	7	600	85	0.85	—	25000	1 Year
SYSKA	SSK-LB	3	210	70	>.95	>85	20000	2 Years
	SSK-QA0301	3	250	83	—	>80	20000	2 Years
	SSK-LB	5	350	70	>.95	>85	20000	2 Years
	SSK-QA0602	5	410	82	—	>80	20000	2 Years
	SSK-LB	7	490	70	>.95	>85	20000	2 Years
	SSK-QA0701	7	600	85	—	>80	20000	2 Years
	SSK-LB	8	640	80	>.95	>85	20000	2 Years
	SSK-QA0901	9	780	86	—	>80	20000	2 Years
	SSK-LB	10	800	80	>.95	>85	20000	2 Years
	SSK-QA0907	10	820	82	—	>80	20000	2 Years
	SSK-LB	12	960	80	>.95	>85	20000	2 Years
	SSK-LB	15	1200	80	>.95	>85	20000	2 Years
	SSK-LB	20	1600	80	>.95	>85	20000	2 Years
Surya	Royal Lamp	3	230	76.67	>0.9	70	20000	2 Years
	Royal Lamp	5	415	83.00	>0.9	70	20000	2 Years
	Royal Lamp	7	680	97.14	>0.9	70	20000	2 Years
	Royal Lamp	9	790	87.78	>0.9	70	20000	2 Years
	Eco Lamp	5	400	80.00	>0.9	70	20000	2 Years
	Eco Lamp	7	560	80.00	>0.9	70	20000	2 Years
	Eco Lamp	9	720	80.00	>0.9	70	20000	2 Years

Crompton	LED-3DFCDL-PRO-BO	3	250	83.33333	> 0.9	NA	5000	2 Years
	LED-5DFCDL-PRO-BO	5	430	86	> 0.9	NA	5000	2 Years
	LED-7DFCDL-PRO-BI	7	600	85.71429	> 0.9	NA	5000	2 Years
	LED-9DFCDL-PRO-BO	9	800	88.88889	> 0.9	NA	5000	2 Years
	LED12DFCDL-PRO-BO	12	1200	100	> 0.9	NA	5000	2 Years
	LED14DFCDL-PRO-BO	14	1400	100	> 0.9	NA	5000	2 Years
	LED18DFCDL-PRO-BO	18	1800	100	> 0.9	NA	5000	2 Years
	LED23WDFCDL-PRO-BO	23	1800	78.26087	> 0.9	NA	5000	2 Years

Table 13: METAL HALIDE LAMPS

Rating–Watts	Life-hours	Initial Lumens
50	15000	3400
70	15000	5600
100	15000	9000
150	10000	13500
175	10000	15000
250	10000	20500
400	20000	36000
1000	12000	110000

Table 14: HIGH PRESSURE SODIUM VAPOUR LAMPS

Rating–Watts	Life-hours	Initial Lumens
35	16000	2250
50	24000	4000
70	24000	5800
100	24000	9500
150	24000	16000
250	24000	27500
400	24000	47500
1000	24000	140000

Table 15: LOW PRESSURE SODIUM VAPOUR LAMPS

Rating-Watts	Life -hours	Initial Lumens
18	18000	1800
35	18000	4550
55	18000	7800
90	18000	14300
135	18000	22600
180	18000	32000

Table 16: PULSE START METAL HALIDE LAMPS

Rating–Watts	Life-hours	Initial Lumens
50	15000	3400
70	15000	5600
100	15000	9000
150	15000	15000
175	15000	17500
200	15000	21000
250	15000	26300
320	20000	34000
400	20000	44000
450	20000	50000

Table 17: COMPARISON OF LIGHT SOURCES

Characteristics	Incandescent	Quartz halogen	Mercury vapour	Metal halide	Pulse start MH	HP SV	LP SV	Compact fluorescent lamp
Initial cost	Low	Low	High	High	High	High	High	High
Initial lumen per watt	15-24	15-24	30-6-	69-115	60-111	79-130	62-150	60-80
Efficiency	poor	poor	fair	good	best	better	best	fair
Average lamp life-hours	1000	2000	24000	15000	30000	24000	20000	10000
Annual operating cost	high	medium	low	low	low	low	very low	low
Long burning hours	fair	fair	good	good	good	good	good	good
Short burning hours	good	good	good	good	good	good	good	good
Colour acceptability	good	very good	fair	very good	very good	fair	fair	good
Degree of light control	very good	very good	fair	good	good	good	poor	fair
Maintained lumen output	good	very good	poor	good	good	good	very good	good
Hot restart time-minutes	instant	instant	10	10-15	2-4	1-3	1	instant

Table 18: PERCENTAGE EFFECTIVE CEILING OR FLOOR REFLECTANCES

(Source: IES lighting handbook-1981)

EFFECTIVE CAVITY REFLECTANCE

BASE REFLECTANCE PER CENT: 90

ROOM CAVITY RATIO \ WALL REFLECTANCE PER CENT	90	80	70	60	50	40	30	20	10	0
0.2	89	88	88	87	86	85	85	84	84	82
0.4	88	87	86	85	84	83	81	80	79	76
0.6	87	86	84	82	80	79	77	76	74	73
0.8	87	85	82	80	77	75	73	71	69	67
1	86	83	80	77	75	72	69	66	64	62
1.2	85	82	78	75	72	69	65	63	60	57
1.4	85	80	77	73	69	65	62	59	57	52
1.6	84	78	75	71	67	63	59	56	53	50
1.8	83	78	73	69	64	60	56	53	50	48
2	83	77	72	67	62	56	53	50	47	43
2.2	82	76	70	65	59	54	50	47	44	40
2.4	82	75	68	64	58	53	48	45	41	37
2.6	81	74	67	62	56	51	46	42	38	35
2.8	81	73	66	60	54	49	44	40	36	34
3	80	72	64	58	52	47	42	38	34	30
3.2	78	71	63	56	50	45	40	36	32	28
3.4	78	70	62	54	48	43	38	34	30	27
3.6	78	69	61	53	47	42	36	32	28	25
3.8	78	69	60	51	45	40	35	31	27	23
4	77	69	58	51	44	39	33	29	25	22
4.2	77	62	57	50	43	37	32	28	24	21
4.4	76	61	56	49	42	36	31	27	23	20
4.6	76	60	55	47	40	35	30	26	22	19
4.8	75	59	54	46	38	34	28	25	21	18
5	75	59	53	45	38	33	28	24	20	16
6	73	61	48	41	34	28	24	20	16	11
7	70	58	45	38	30	27	21	18	14	8
8	68	55	42	35	27	23	18	15	12	6
9	66	52	38	31	25	21	16	14	11	5
10	65	51	36	29	22	19	15	11	9	4

BASE REFLECTANCE PER CENT: 80

ROOM CAVITY RATIO \ WALL REFLECTANCE PER CENT	90	80	70	60	50	40	30	20	10	0
0.2	79	78	78	77	77	76	76	75	74	72
0.4	79	77	76	75	74	73	72	71	70	68
0.6	78	76	75	73	71	70	68	66	65	63
0.8	78	75	73	71	69	67	65	63	61	57
1	77	74	72	69	67	65	62	60	57	55
1.2	76	73	70	67	64	61	58	55	53	51
1.4	76	72	68	65	62	59	55	53	50	48
1.6	75	71	67	63	60	57	53	50	47	44
1.8	75	70	66	62	58	54	50	47	44	41
2	74	69	64	60	56	52	48	45	41	38
2.2	74	68	63	58	54	49	45	42	38	35
2.4	73	67	61	56	52	47	43	40	36	33
2.6	73	66	60	55	50	45	41	38	34	31
2.8	73	65	59	53	48	43	39	36	32	29
3	72	65	58	52	47	42	37	34	30	27
3.2	72	65	37	51	45	40	35	33	28	25
3.4	71	64	56	49	44	38	34	32	27	24
3.6	71	63	54	48	43	38	32	30	25	23
3.8	70	62	53	47	41	36	31	28	24	22
4	70	61	53	46	40	35	30	26	22	20
4.2	69	60	52	45	39	34	29	25	21	18
4.4	69	60	51	44	38	33	28	24	20	17
4.6	69	59	50	43	37	32	27	23	19	15
4.8	68	58	49	42	36	31	26	22	18	14
5	68	58	48	41	35	30	25	21	18	14
6	66	55	44	38	31	27	22	19	15	10
7	64	53	41	35	28	24	19	16	12	7
8	62	50	38	32	25	21	17	14	11	5
9	61	48	36	30	23	19	15	13	10	4
10	59	46	33	27	21	18	14	11	8	3

BASE REFLECTANCE PER CENT: 70

ROOM CAVITY RATIO \ WALL REFLECTANCE PER CENT	90	80	70	60	50	40	30	20	10	0
0.2	70	69	68	68	67	67	66	66	65	64
0.4	69	68	67	66	65	64	63	62	61	58
0.6	69	67	65	64	63	61	59	58	57	54
0.8	68	66	64	62	60	58	56	55	53	50
1	68	65	62	60	58	55	53	52	50	47
1.2	67	64	61	59	57	54	50	48	46	44
1.4	67	63	60	58	55	51	47	45	44	41
1.6	67	62	59	56	53	47	45	43	41	38
1.8	66	61	58	54	51	46	42	40	38	35
2	66	60	56	52	49	45	40	38	36	33
2.2	65	60	55	51	48	43	38	36	34	32
2.4	65	60	54	50	46	41	37	35	32	30
2.6	65	59	54	49	45	40	35	33	30	28
2.8	65	59	53	48	43	38	33	30	28	26
3	66	58	52	47	42	37	32	29	27	24
3.2	64	58	51	46	40	36	31	28	25	23
3.4	64	57	50	45	39	35	29	27	24	22
3.6	63	56	49	44	38	33	28	25	22	20
3.8	63	56	49	43	37	32	27	24	21	19
4	63	55	48	42	36	31	26	23	20	17
4.2	62	55	47	41	35	30	25	22	19	16
4.4	62	54	46	40	34	29	24	21	18	15
4.6	62	53	45	39	33	28	24	21	17	14
4.8	62	53	45	38	32	27	23	20	16	13
5	61	52	44	36	31	26	22	19	16	12
6	60	51	41	35	28	24	19	16	13	9
7	58	48	38	32	26	22	17	14	11	6
8	57	46	35	29	23	19	15	13	10	5
9	56	45	33	27	21	18	14	12	9	4
10	55	43	31	25	19	16	12	10	8	3

BASE REFLECTANCE PER CENT: 60

ROOM CAVITY RATIO \ WALL REFLECTANCE PER CENT	90	80	70	60	50	40	30	20	10	0
0.2	60	59	59	59	58	57	56	56	55	53
0.4	60	59	59	58	57	55	54	53	52	50
0.6	60	58	57	56	55	53	51	51	50	46
0.8	59	57	56	55	54	51	48	47	46	43
1	59	57	55	53	51	48	45	44	43	41
1.2	59	56	54	51	49	46	44	42	40	38
1.4	59	56	53	49	47	44	41	39	38	36
1.6	59	55	52	48	45	42	39	37	35	33
1.8	58	55	51	47	44	40	37	35	33	31
2	58	54	50	46	43	39	35	33	31	29
2.2	58	53	49	45	42	37	34	31	29	28
2.4	58	53	48	44	41	36	32	30	27	26
2.6	58	53	48	43	39	35	31	28	26	24
2.8	58	53	47	43	38	34	29	27	24	22
3	57	52	46	42	37	32	28	25	23	20
3.2	57	51	45	41	36	31	27	23	22	18
3.4	57	51	45	40	35	30	26	23	20	17
3.6	57	50	44	39	34	29	25	22	19	16
3.8	57	50	43	38	33	29	24	21	19	15
4	57	49	42	37	32	28	23	20	18	14
4.2	56	49	42	37	32	27	22	19	17	14
4.4	56	49	42	36	31	27	22	19	16	13
4.6	56	49	41	35	30	26	21	18	16	13
4.8	56	48	41	34	29	25	21	18	15	12
5	56	48	40	34	28	24	20	17	14	11
6	55	45	37	31	25	21	17	14	11	7
7	54	43	35	30	24	20	15	12	9	5
8	53	42	33	28	22	18	14	11	8	4
9	52	40	31	26	20	16	12	10	7	3
10	51	39	29	24	18	15	11	9	7	2

BASE REFLECTANCE PER CENT: 50

ROOM CAVITY RATIO \ WALL REFLECTANCE PER CENT	90	80	70	60	50	40	30	20	10	0
0.2	50	50	49	49	48	48	47	46	46	44
0.4	50	49	48	48	47	46	45	45	44	42
0.6	50	48	47	46	45	44	43	42	41	38
0.8	50	48	47	45	44	42	40	38	38	36
1	50	48	46	44	43	41	38	37	36	34
1.2	50	47	45	43	41	39	36	35	34	29
1.4	50	47	45	42	40	38	35	34	32	27
1.6	50	47	44	41	39	36	33	32	30	26
1.8	50	46	43	40	38	35	31	30	28	25
2	50	46	43	40	37	34	30	28	26	24
2.2	50	46	42	38	36	33	29	27	24	22
2.4	50	46	42	37	35	31	27	25	23	21
2.6	50	46	41	37	34	30	26	23	21	20
2.8	50	46	41	36	33	29	25	22	20	19
3	50	45	40	36	32	28	24	21	19	17
3.2	50	44	39	35	31	27	23	20	18	16
3.4	50	44	39	35	30	26	22	19	17	15
3.6	50	44	39	34	29	25	21	18	16	14
3.8	50	44	38	34	29	25	21	17	15	13
4	50	44	38	33	28	24	20	17	15	12
4.2	50	43	37	32	28	24	20	17	14	12
4.4	50	43	37	32	27	23	19	16	13	11
4.6	50	43	36	31	26	22	18	15	13	10
4.8	50	43	36	31	26	22	18	15	12	9
5	50	42	35	30	25	21	17	14	12	9
6	50	42	34	29	23	19	15	13	10	6
7	49	41	32	27	21	18	14	11	8	5
8	49	40	30	25	19	16	12	10	7	3
9	48	39	29	24	18	15	11	9	7	3
10	47	37	27	22	17	14	10	8	6	2

BASE REFLECTANCE PER CENT	40										30										20									
WALL REFLECTANCE PER CENT	90	80	70	60	50	40	30	20	10	0	90	80	70	60	50	40	30	20	10	0	90	80	70	60	50	40	30	20	10	0
ROOM CAVITY RATIO																														
0.2	40	40	39	39	39	38	38	37	36	36	31	31	30	30	29	29	29	28	28	27	21	20	20	20	20	20	19	19	19	17
0.4	41	40	39	39	38	37	36	35	34	34	31	31	30	30	29	28	28	27	26	25	22	21	20	20	20	19	19	18	18	16
0.6	41	40	39	38	37	36	34	33	32	31	32	31	30	29	28	27	26	26	25	23	23	21	21	20	19	19	18	18	17	15
0.8	41	40	38	37	36	35	33	32	31	29	32	31	30	29	28	26	25	25	23	22	24	22	21	20	19	19	18	17	16	14
1	42	42	38	37	35	33	32	31	29	27	33	32	30	29	27	25	24	23	22	20	25	23	22	20	19	18	17	16	15	13
1.2	42	40	38	36	34	32	30	29	27	25	33	32	30	28	27	25	23	22	21	19	25	23	22	20	19	17	17	16	14	12
1.4	42	39	37	35	33	31	29	27	25	23	34	32	30	28	26	24	22	21	19	18	26	24	22	20	18	17	16	15	13	12
1.6	42	39	37	35	32	30	27	25	23	22	34	33	29	27	25	23	22	20	18	17	26	24	22	20	18	17	16	15	13	11
1.8	42	39	36	34	31	29	26	24	22	21	35	33	29	27	25	23	21	19	17	16	27	25	23	20	18	17	15	14	12	10
2	42	39	36	34	31	28	25	23	21	19	35	33	29	26	24	22	20	18	16	14	28	25	23	20	18	16	15	13	11	9
2.2	42	39	36	33	30	27	24	22	19	18	36	32	29	26	24	22	19	17	15	13	28	25	23	20	18	16	14	12	10	9
2.4	43	39	35	33	29	27	24	21	18	17	36	32	29	26	24	22	19	16	14	12	29	26	23	20	18	16	14	12	10	8
2.6	43	39	35	32	29	26	23	20	17	15	36	32	29	25	23	21	18	16	14	12	29	26	23	20	18	16	14	11	9	8
2.8	43	39	35	32	28	25	22	19	16	14	37	33	29	25	23	21	17	15	13	11	30	27	23	20	18	15	13	11	9	7
3	43	39	35	31	27	24	21	18	16	13	37	33	29	25	22	20	17	15	12	10	30	27	23	20	17	15	13	11	9	7
3.2	43	39	35	31	27	23	20	17	15	13	37	33	29	25	22	19	16	14	12	10	31	27	23	20	17	15	12	11	9	6
3.4	43	39	34	30	26	23	20	17	14	12	37	33	29	25	22	19	16	14	11	9	31	27	23	20	17	15	12	10	8	6
3.6	44	39	34	30	26	22	19	16	14	11	38	33	29	24	21	18	15	13	10	9	32	27	23	20	17	15	12	10	8	5
3.8	44	38	33	29	25	22	18	16	13	10	38	33	28	24	21	18	15	13	10	8	32	28	23	20	17	15	12	10	7	5
4	44	38	32	29	25	21	18	15	12	10	38	33	28	24	21	18	14	12	9	7	33	28	22	20	17	14	11	9	7	5
4.2	44	38	33	29	24	21	17	15	12	10	38	33	28	24	20	17	14	12	9	7	33	28	23	20	17	14	11	9	7	4
4.4	46	38	33	28	24	20	17	14	11	9	39	33	28	24	20	17	14	11	9	6	34	28	24	20	17	14	11	9	7	4
4.6	44	38	32	28	23	19	16	14	11	8	39	33	28	24	20	17	13	10	8	6	34	29	24	20	17	14	11	9	7	4
4.8	44	38	32	27	22	19	16	13	10	8	39	33	28	24	20	17	13	10	8	5	35	29	24	20	17	13	10	8	6	4
5	45	38	31	27	22	19	15	13	10	7	39	33	28	24	19	16	13	10	8	5	35	29	24	20	16	13	10	8	6	4
6	44	37	30	25	20	17	13	11	8	5	39	33	27	23	18	15	11	9	6	4	36	30	24	20	16	13	10	8	5	2
7	44	36	29	24	19	16	12	10	7	4	40	33	26	22	17	14	10	8	5	3	36	30	24	20	15	12	9	7	4	2
8	44	35	28	23	18	15	11	9	6	3	40	33	26	21	16	13	9	7	4	2	37	30	23	19	15	12	8	6	3	1
9	44	35	26	21	16	13	10	8	5	2	40	33	25	20	15	12	9	7	4	2	37	29	23	19	14	11	8	6	3	1
10	43	34	25	20	15	12	8	7	5	2	40	32	24	19	14	11	8	6	3	1	37	29	23	18	13	10	7	5	3	1

BASE REFLECTANCE PER CENT	10										0									
WALL REFLECTANCE PER CENT	90	80	70	60	50	40	30	20	10	0	90	80	70	60	50	40	30	20	10	0
ROOM CAVITY RATIO																				
0.2	11	11	11	10	10	10	10	9	9	9	2	2	2	1	1	1	1	0	0	0
0.4	12	11	11	11	11	10	10	9	9	8	4	3	3	2	2	2	1	1	0	0
0.6	13	13	12	11	11	10	10	9	8	8	5	5	4	3	3	2	2	1	1	0
0.8	15	14	13	12	11	10	10	9	8	7	7	6	5	4	4	3	2	2	1	0
1	16	14	13	12	12	11	10	9	8	7	8	7	6	5	4	3	2	2	1	0
1.2	17	15	14	13	12	11	10	9	7	6	10	8	7	6	5	4	3	2	1	0
1.4	18	16	14	13	12	11	10	9	7	6	11	9	8	7	6	4	3	2	1	0
1.6	19	17	15	14	12	11	9	8	7	6	12	10	9	7	6	5	3	2	1	0
1.8	19	17	15	14	13	11	9	8	6	5	13	11	9	8	7	5	4	3	1	0
2	20	18	16	14	13	11	9	8	6	5	14	12	10	9	7	5	4	3	1	0
2.2	21	19	16	14	13	11	9	7	6	5	15	13	11	9	7	6	4	3	1	0
2.4	22	19	17	15	13	11	9	7	6	5	16	13	11	9	8	6	4	3	1	0
2.6	23	20	17	15	13	11	9	7	6	4	17	14	12	10	8	6	5	3	2	0
2.8	23	20	18	16	13	11	9	7	5	3	17	15	13	10	8	7	5	3	2	0
3	24	21	18	16	13	11	9	7	5	3	18	16	13	11	9	7	5	3	2	0
3.2	25	21	18	16	13	11	9	7	5	3	19	16	14	11	9	7	5	3	2	0
3.4	26	22	18	16	13	11	9	7	5	3	20	17	14	12	9	7	5	3	2	0
3.6	26	22	19	16	13	11	9	6	4	3	20	17	15	12	10	8	5	4	2	0
3.8	27	23	19	17	14	11	9	6	4	2	21	18	15	12	10	8	5	4	2	0
4	27	23	20	17	14	11	9	6	4	2	22	18	15	13	10	8	5	4	2	0
4.2	28	24	20	17	14	11	8	6	4	2	22	19	16	13	10	8	6	4	2	0
4.4	28	24	20	17	14	11	8	6	4	2	23	19	16	13	10	8	6	4	2	0
4.6	29	25	20	17	14	11	8	6	4	2	23	20	17	13	11	8	6	4	2	0
4.8	29	25	20	17	14	11	8	6	4	2	24	20	17	14	11	8	6	4	2	0
5	30	25	20	17	14	11	8	6	4	2	25	21	17	14	11	8	6	4	2	0
6	31	26	21	18	14	11	8	6	3	1	27	23	18	15	12	9	6	4	2	0
7	32	27	21	17	13	11	8	6	3	1	28	24	19	15	12	9	6	4	2	0
8	33	27	21	17	13	10	7	5	3	1	30	25	20	15	12	9	6	4	2	0
9	34	28	21	17	13	10	7	5	2	1	31	25	20	15	12	9	6	4	2	0
10	34	28	21	17	12	10	7	5	2	1	31	25	20	15	12	9	6	4	2	0

Table 19: FLEXIBLE HOUSE WIRING SINGLE CORE CABLES

(With copper conductor) 250/440 V

Conductor nominal area (mm^2)	No. & diameter of wire	Resistance (Ω/km)	Current carrying capacity (A)	Thickness (mm)		Overall diameter (mm)		Equivalent SWG
1.0	14.0/0.3	18.89	11	0.9	0.65	3.00	4.55	1/18
1.6	23.0/0.3	11.50	15	0.9	0.65	3.50	4.80	3/22
2.0	28.0/0.3	9.42	17	0.9	0.65	3.90	5.40	3/20
3.0	44.0/0.3	6.00	22	1.0	0.75	4.15	5.55	7/22
4.5	65.0/0.3	4.06	26	1.0	0.75	4.90	6.10	7/20

Colours

Single core unsheathed – Red, Yellow, Blue, Black, Green, and White.
Single core sheathed. Core colour black. Sheath White.
Packing: 91.44 m in convenient to handle corrugated boxes.

Table 20: COMPARATIVE PROPERTIES OF PVC WIRING CABLES (Source: Reliance Cables)

Feature	Normal PVC	Heat resistant HR PVC	Fire retardant FR PVC	Fire retardant- Low smoke FRLS PVC	Zero halogen Low smoke ZHLS
Insulation material	PVC	PVC	Special PVC	Special PVC	Special polymer
Insulation property	Normal	Good	Good	Good	Very Good
Temperature rating	70°C	85°C	70°C	70°C	85°C
Thermal stability	Normal	Very good	Good	Good	Very good
Flame retardant property	Good	Good	Very good	Very good	Excellent
Safety during burning	Average	Average	Good	Good	Excellent
Requirement of oxygen to catch fire (% of air)	> 21	> 21	> 30	> 30	> 35
Temperature required to catch fire (with 21% oxygen)	Room temperature	Room temperature	> 250°C	> 250°C	> 300°C
Visibility during cable burning (%)	< 20	< 20	< 35	> 40	> 80
Release of halogen gas during burning (% by weight)	< 20	< 20	< 20	< 20	zero
Abrasion resistance during installation	Good	Good	Good	Good	Good

Table 21: RECOMMENDED RATINGS OF LIGHT AND SOCKET POINTS (NEC-2011)

Specification	Rating (watts)
Incandescent lamps	60
Fans (ceiling)	60
Table fan	60
6-A three pin sockets	100
16-A three pin sockets	1000
Fluorescent lamp 600 mm	25
Fluorescent lamp 1200 mm	50
Fluorescent lamp 1500 mm	90
Exhaust fan 300 mm	45

Table 22: RECOMMENDED NUMBER OF POINTS IN DWELLING UNITS (NEC-2011)

Description	Area of the residential building (m^2)				
	35	45	55	85	140
Light points	7	8	10	12	17
Ceiling fans*	2-2	3-2	4-3	5-4	7-5
6A sockets	2	3	4	5	7
16A sockets	-	1	2	3	4
Call bell	-	-	1	1	1

Table 23: DETAILS OF SOCKET OUTLETS REQUIRED FOR DOMESTIC BUILDINGS (NEC-2011)

Description	Socket outlets	
	6A	16A
Bedrooms	2-3	1
Living rooms	2-3	2
Kitchen	1	2
Dining room	2	1
Garage	1	1
For refrigerator		1
For air-conditioner		1
Verandah	1 for every 10 m^2	
Bathrooms	1	1

Table 24: DIFFERENT TYPES OF MCB DISTRIBUTION BOARDS
(Source: M/s Legrand India)

Type description	No. of phases	Protection class	Number of ways	No. of MCB modules
Consumer units conforming to IS-8623, Universal mounting with 100 A copper busbar, neutral bar, earth bus, cable ties and removable gland plates at top and bottom. Provision for double pole RCD/MCB/Isolator as incoming and single pole MCB as outgoing	Single phase	IP 20 & IP 42 with metal or acrylic door	4 8 12	4 + 4 4 + 8 4 + 12
Regular DBs conforming to IS 8623 with 100A busbars, neutral and earth busbars.	Single phase	IP 20	4 8 12	4 8 12
Flush mounting DBs conforming to IS-8623 and with reversible pan assembly for top and bottom cable entry, Masking sheet for protection against entry of concrete, and with provision for double pole RCCB/MCB/Isolator as incoming and MCBs as outgoing	Single phase	IP 20	4 8 12	2 + 4 2 + 8 2 + 12
Consumer units conforming to IS- 8623, Universal mounting with 100 A copper busbar, neutral bar, earth bus, cable ties and removable gland plates at top and bottom. Provision for four pole RCCB as incoming and three single pole MCBs as outgoing	Three phase	IP 20 & IP 42 with metal or acrylic door	3 × 3	6 + 9
ETPN DBs conforming to IS-8623, universal mounting, with 100A copper busbars, neutral and earth bars, cable ties, top and bottom removable gland plates with knock outs and suitable for fixing four pole RCD/MCB/Isolator as incoming and single pole MCB as outgoing.	Three phase	IP 20 & IP 42 with metal or acrylic doors	4 6 8' 12	8 + 12 8 + 18 8 + 24 8 + 36

Per phase isolation DBs conforming to IS-8623, universal mounting, with 100A copper busbars, neutral and earth bars, cable ties, top and bottom removable gland plates with knock outs and suitable for fixing four pole RCD/MCB/Isolator as incoming, 3 double pole RCD/MCB/Isolator as sub incomer and single pole MCB as out going with separate neutral bars. (The advantage of per phase isolation DBs are that only the phase where there is an earth leakage fault is isolated and the other two phases are in operation, avoiding total darkness).	Three phase	IP 20 & IP 42 with metal or acrylic doors	4 6 8 12	8 + 12 8 + 18 8 + 24 8 + 36
Phase selector DBs as per IS 8623, universal mounting with 100 A copper busbar, neutral and earth bar, top and bottom removable gland plates with provision for 3 Nos. 63A selector switches, 3 Nos. piano type indicating lamp switch, four pole RCD/MCB/Isolator as incoming and SP MCBs as outgoing.	Three phase	IP 20 & IP 42 with metal doors	4 6 8 12	6 + 12 6 + 18 6 + 24 6 + 36
Seven segment DBs as per IS 8623, with 100 A copper busbar, neutral and earth bar, top and bottom removable gland plates with provision for pole RCD/MCB/Isolator as incoming, DP RCDs/MCBs/Isolator as sub incomer and SP MCBs as outgoing. (These DBs provide phase segregation and separation between incomer and outgoing)	Three phase	IP 20 & IP 42 with metal doors	4 6 8 12	4 + 12 + 12 4 + 12 + 18 4 + 12 + 24 4 + 12 + 36
VTPN DBs as per IS 8623, universal mounting with 200 A copper busbars, earth and neutral bars, cable ties, top and bottom removable gland plates, and with provision for four pole RCD or RCBO/MCB/Isolator as incoming and SP/TP MCBs as outgoing. (The advantage of this DB is that both single and three phase outgoings can be taken from this DB)	Three phase	IP 20 & IP 42 with metal or acrylic doors	4 8 12	8 + 12 8 + 24 8 + 36

Table 25: MAXIMUM PERMISSIBLE NUMBER OF 1.1 kV GRADE CABLES IN METALLIC CONDUITS (NEC-2011)

Size of cable mm^2	Size of conduits –(mm)				
	20	**25**	**32**	**40**	**50**
1	7	13	20		
1.5	6	10	14		
2.5	5	6	10	14	
4	3	6	10	11	
6	2	5	9	11	
10		4	7	9	
16		2	4	5	12
25			2	2	6

Table 26: PERFORMANCE DATA OF TYPICAL EXHAUST FANS

Fan sweep (mm)	Fan speed (RPM)	Power input (W)	Sound level (dB)	Air displacement (Free air flow condition) CMH
230	1370	40	49	760
300	920	45	46	1270
300	1420	82	FQ	1920
380	920	78	Q	2460
380	1420*	150	NI	4000
450	720	90	Q	3400
450	920	132	FQ	4340
450	1420*	370	FN	6800
600	720	240	FQ	7900
600	920*	500	NI	10450

* Not recommended to be used with louvre shutters

CMH Cubic metre per hour

Q Quiet (50-55 dB)

FQ Fairly quiet (55-60 dB)

NI Normal industrial noise (60-65 dB)

FN Fairly noisy (Above 65 dB)

Table 27: PERFORMANCE DATA FOR CEILINGS FANS (Source: Crompton Greaves)

Type	Sweep mm	Speed rpm	Power W	Air delivery m^3/min
High speed	600	900	64	120
	900	390	60	155
	1050	370	66	195
	1200	350	70	230
	1400	290	76	270
High breeze	1200	310	70	230
	1400	275	74	270
Cool breeze	1050	420	69	160
	1200	380	80	220
	1400	290	85	270
Power saver	1200	330	50	220

Table 28: SIZEOF CABLES AND EARTHING CONDUCTORS FOR MOTORS

Motor rating		Approx. full load current	Backup fuse rating		Cable size AYFY	Earth Conductor size		
						Copper		Al
kW	HP	Ampere	DOL	Assisted	mm^2	mm^2	SWG	mm^2
0.75	1	1.8	6	-	4	3.251	10	25
1.50	2	3.5	10	-	4	3.251	10	25
2.25	3	5.0	16	-	4	3.251	10	25
3.75	5	7.5	25	-	4	3.251	10	25
5.50	7.5	11.0	32	25	6	3.251	10	25
7.5	10	14.0	32	25	6	3.251	10	25
9.3	12.5	18.0	50	35	10	3.251	10	25
11.2	15	21.0	50	35	10	3.251	10	25
15.0	20	28.0	63	50	16	3.251	10	25
18.7	25	36.0	80	63	25	3.251	10	25
22.5	30	40.0	100	63	25	4.877	6	35
26.0	35	47.0	120	100	35	4.877	6	35
30.0	40	55.0	120	100	35	4.877	6	35
37.5	50	66.0	160	100	50	4.877	6	35
45.0	60	80.0	200	120	70	5.893	4	50
55.0	75	95.0	200	120	95	5.893	4	50
59.0	80	105.0	250	200	150	6.910	2	70
67.5	90	118.0	250	200	185	6.910	2	70

75.0	100	135.0	250	200	185	6.910	2	70
90.0	125	165.0	320	250	240	6.910	2	95
110.0	150	200.0	350	320	400	7.620	1 or 0	120
131.0	175	230.0	400	320	2×150	7.620	1 or 0	120
150.0	200	270.0	500	350	2×185	7.620	1 or 0	120
185.0	250	3250.0	600	500	2×240	25×3		185

For motors of rating 185 kW and above, breaker shall be provided and the following protection shall also be incorporated.

(a) Thermal O/L
(b) Short circuit
(c) Earth fault
(d) Negative sequence
(e) Under voltage
(f) Locked rotor

Table 29: TYPICAL IMPEDANCE VALUES OF 11 kV/433 V DISTRIBUTION TRANSFORMERS

Rated power (kVA)	Oil immersed transformers				Cast resin transformers			
	Impedance voltage (%)	R (mΩ)	X (mΩ)	Z (mΩ)	Impedance voltage (%)	R (mΩ)	X (mΩ)	Z (mΩ)
100	4	37.9	59.5	70.6	6	37.0	99.1	105.8
160	4	16.2	41.0	44.1	6	18.6	63.5	66.2
200	4	11.9	33.2	35.3	6	14.1	51.0	52.9
250	4	9.2	26.7	28.2	6	10.7	41.0	42.3
315	4	6.2	21.5	22.4	6	8.0	32.6	33.6
400	4	5.1	16.9	17.6	6	6.1	25.8	26.5
500	4	3.8	13.6	14.1	6	4.6	20.7	21.2
630	4	2.9	12.9	13.2	6	3.5	16.4	16.8
800	6	2.9	10.8	11.2	6	2.6	13.0	13.2
1000	6	2.3	10.3	10.6	6	1.9	10.4	10.6
1250	6	1.8	8.3	8.5	6	1.5	8.3	8.5
1600	6	1.4	6.5	6.6	6	1.1	6.5	6.6
2000	6	1.1	5.2	5.3	6	0.9	5.2	5.3

Table 30: MAXIMUM RATING OF A MOTOR THAT CAN BE CONNECTED TO A TRANSFORMER

Method of starting	Starting current of motors in multiples of full load current	Motor rating permissible in HP
DOL	7.0	(2S-D)/7.0
ATS with 80% tap	5.5	(2S-D)/5.5
ATS with 40% tap, Star Delta	2.5	(2S-D)/2.5
Soft starter	2.0	(2S-D)/2.0
Slip ring	2.0	(2S-D)/2.0

Where, S is the rating of transformer in kVA and D the maximum demand excluding the motor in question. When the number of starts/hour is more than 6 or loading cycle is more than 6, maximum rating shall be limited to 2% of the fault level.

Table 31: CABLES ON SECONDARY SIDE OF TRANSFORMER (11 kV/433 V)

Capacity of Transformer (kVA)	Primary current (A)	Secondary current (A)	Minimum size of neutral earthing conductor (mm × mm)	Minimum size of AYFY cable (mm^2)
63	3.3	84	25×3	50
100	5.25	133.3	25 × 3	95 (2 × 50)
160	8.40	213.3	25 × 3	185 (2 × 95)
200	10.49	266.6	25 × 3	300 (2 × 120)
250	13.12	333	25 × 3	(2 × 185)
315	16.53	420	31 × 3 or 25 × 4	(2 × 300) or
400	21.80	533	38 × 3	(3 × 185)
500	26.20	666.5	25 × 6	(3 × 300) or
630	33	840	31 × 6	(2 × 400)
750	39.36	1000	50 × 4	(3 × 400) or
1000	52.50	1333	210 mm^2	(4 × 240)
1250	65.60	1667	290 mm^2	(4 × 400)
1600	83.98	2133	380 mm^2	Busbar trunking
2000	105.00	3666	450 mm^2	Busbar trunking

Above sizes are recommended assuming following de-rating factors: Enclosure effect – 0.75, Temperature effect – 0.88, and Impurity effect – 0.94.

Busbar trunking shall be designed for nominal minimum fault current of 25 kA or 50 kA depending on the actual fault level. 50 kA short circuit current rating is recommended for all short circuit currents above 25 kA.

Maximum distance between bus supports shall be 50 cm with central stiffeners and bolting for phase busbar. S.M.C. finger type 9 mm thick single supports are recommended for 25 kA rating and double supports or 20 mm thick single supports or 20 mm thick single supports are recommended for 50 kA rating. Ventilation louvres shall be provided as an anti-condensate measure for the enclosure.

Table 31A: STANDARD FEEDER RATING, CABLE SIZE AND RECOMMENDED BACK UP FUSE RATINGS

Feeder current rating (A)	Feeder cable size AYWY/ AYFY (mm^2)	Recommended fuse rating/ MCCB rating (A)	Recommended switch rating for SFUs (A)
32	4 × 10	32	32
40	4 × 16	40	63
50	4 × 25	50	63
63	3.5 × 35	63	63
80	3.5 × 50	80	100
100	3.5 × 70	100	100
125	3.5 × 120	125	125
160	3.5 × 185	160	160
200	3.5 × 240	200	200
250	3.5 × 400	250	250
320	2 runs of 3.5 × 185	315	315
400	2 runs of 3.5 × 240	400	400
500	3 runs of 3.5 × 240	500	500
630	3 runs of 3.5 × 400	630	630
800	4 runs of 3.5 × 400	800	800

GUIDELINES FOR SWITCHBOARDS IN MV INSTALLATIONS

The design of industrial switchboards is a complex process taking into consideration the magnitude of possibilities that exist with the different switching combinations. Basic guidelines of the design of MV switchboards are listed below.

1 Busbar section shall be selected based on a current density of 1.2A/mm^2 for copper and 0.8 A/mm^2 for aluminium. The minimum thickness of busbar shall be 6 mm.

2 The combinations of switchgears is used in switchboards shown in the table below.

Table 31B: SWITCHGEAR COMBINATIONS OF MV SWITCHBOARDS

Incomer switchgear	Outgoing switchgear
ACB	SFU, MCCB or MCB
MCCB	SFU, MCCB, MCB
SFU	MCCB, MCB, SFU
Isolator	MCCB, MCB, SFU

3 When ACB is used at the incoming side of a switchboard, the outlet switch rating shall not be less than 1/3 of the incomer setting of the ACB.

4 When MCCB is provided at the incomer, the outgoing switch rating shall not be less than 1/5 of the incomer MCCB setting.

5 When CB is provided at the incomer and SFU at the outgoing, outgoing fuse rating shall not exceed 1/3 of the incomer breaker setting.

6 For ACB as incomer and MCCB as outgoing, MCCBs can be set at 80% of incomer breaker setting.

7 If MCCBs are provided at incomer and outgoing, maximum current setting of outgoing MCCBs shall not exceed 50% of incomer MCCB setting.

8 In case of breaker controlled switchboards (3), even though the outlet switch rating is fixed at 1/3 incomer setting, the feeder can be derated by using lower size fuses.

9 For SFU/SFU controlled switchboards, the outgoing SFU can be 1/10 of incomer SFU rating. Maximum fuse rating at outlet shall not exceed 1/2 the incomer fuse rating.

10 Maximum rating of fuses shall be 630 A and maximum rating of isolator shall be 800 A.

11 Maximum incomer rating of a LDB shall be 63 A. The maximum number of ways of the DB shall be 12.

12 The maximum capacity of a motor that can be connected to a DB shall be less than 7.5 kW.

13 VDBs can use switchgears up to 100 A as incomer.

14 Separate feeders shall be used for all motors above 7.5 kW.

15 The minimum clearance in front of a switchboard shall be 1 m.

16 Minimum depth of laying of cables shall be 75 cm and at road crossings 1m and drawn through 100 mm diameter hume pipes.

17 Minimum size of cable for power loads shall be 4 mm^2 Al and 2.5 mm^2 copper.

18 Voltage drop in cable shall not exceed 3% of the rated voltage.

19 Cables shall be loaded only up to 75% of the rated current capacity as specified by the manufacturer.

20 All cables should have adequate short circuit current carrying capacity.

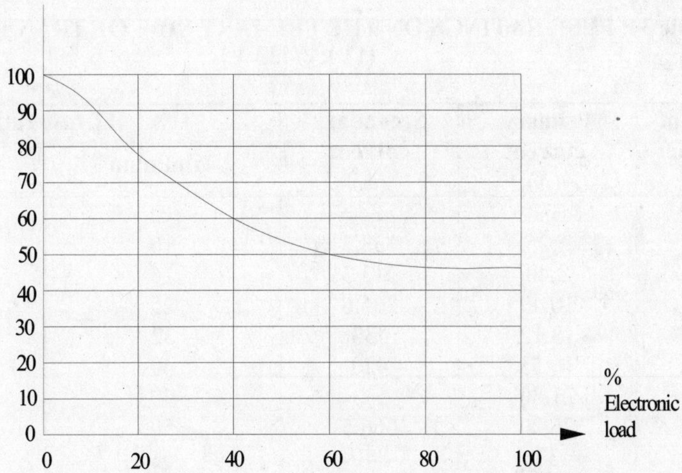

Fig. 32. PERCENTAGE DE-RATING REQUIRED FOR A TRANSFORMER SUPPLYING ELECTRONIC (NONLINEAR) LOADS

Table 33: RECOMMENDED SIZES OF ALUMINIUM BUSBARS FOR BUS TRUNKING

Transformer rating (kVA)	Current (A)	Bus size-Phase (mm)	Bus size-neutral (mm)	Enclosure size (mm)
750	1200	1 × 127 × 12.7 or 2 × 101.6 × 6.35	1 × 127 × 6.35 or 1 × 101.6 × 6.35	600 × 250
1000	1600	2 × 152.4 × 6.35 or 2 × 127 × 9.53	1 × 152.4 × 6.35 or 1 × 127 × 9.53	600 × 300
1250	2000	2 × 152.4 × 12.7 or 4 × 127 × 6.35	1 × 152.4 × 12.7 or 2 × 127 × 6.35	800 × 400
1600	2500	2 × 203.2 × 12.7 or 4 × 152.4 × 6.35	1 × 20.3.2 × 12.7 or 2 × 152.4 × 6.35	800 × 500
2000	3000	4 × 152.4 × 9.53	2 × 152.4 × 9.53	800 × 600

Table 34: FUSE RATINGS ON THE PRIMARY SIDE OF TRANSFORMERS (11 kV/433 V)

Capacity of transformer (kVA)	Primary current (A)	Secondary current (A)	HT fuse rating-A	
			Minimum	Maximum
63	3.3	84	10	16
100	5.25	133.3	16	25
160	8.40	213.3	16	40
200	10.49	266.6	25	40
250	13.12	333	32	40
315	16.53	420	40	63
400	21.80	533	40	63
500	26.20	666.5	50	100
630	33	840	63	100
750	39.36	1000	75	160
1000	52.50	1333	100	160
1250	65.60	1667	100	200
1600	83.98	2133	160	250
2000	105.00	2666	160	250

Table 35: OVERLOADING CAPACITIES OF OIL IMMERSED TRANSFORMERS

Previous continuous loading as a % of the rating	Overloading duration				
	10%	20%	30%	40%	50%
50	3 hrs	1.5 hrs	60 min	30 min	15 min
75	2 hrs	1.0 hr	30 min	15 min	8 min
90	1 hr	30 min	15 min	8 min	4 min

Table 36: OVERLOADING CAPACITIES OF DRY TYPE TRANSFORMERS

Previous continuous loading as a % of the rating	Overloading duration				
	10%	20%	30%	40%	50%
50	60 min	30 min	20 min	15 min	12 min
75	55 min	23 min	15 min	11 min	9 min
90	45 min	16 min	10 min	7 min	5 min

Table 37: ABSOLUTE TEMPERATURE WITHSTAND CAPACITY OF DIFFERENT INSULATION CLASSES

Insulation class	Winding temperature rise in °C as per IS 2026-part-2	Absolute temperature withstand capacity in °C as per IS 1271
A	50	105
E	65	120
B	70	130
F	90	155
H	115	>180

Table 38A: CURRENT RATING OF ALUMINIUM BUSBAR SECTIONS (Indal D50S WP Rectangular Busbar: Current Ratings)

Size mm	Indal ref. no.	1 Bar	2 Bars	3 Bars	4 Bars
		Ampere (AC)			
25.4×6.35	F204	355	705	970	1100
38.1×6.35	F13	520	1020	1315	1535
50.8×6.35	F15	670	1290	1705	1940
63.5×6.35	F74	810	1510	2000	2260
76.2×6.35	F33	960	1740	2310	2620
101.6×6.35	F36	1235	2140	2800	3200
127.0×6.35	F249	1505	2510	3240	3700
152.4×6.35	271	1780	2860	3680	4240
50.8×9.53	F16	830	1500	1970	2260
76.2×9.53	F182	1180	2050	2660	3030
101.6×9.53	F96	1495	2480	3150	3560
127.0×9.53	F113	1860	2930	3660	4200
152.4×9.53	F272	2120	3340	4080	4680
203.2×9.53	F273	2750	4150	4900	5740
76.2×12.7	F34	1355	2240	2830	3240
101.6×12.7	F66	1740	2720	3360	3900
127.0×12.7	F144	2080	3120	3900	4550
152.4×12.7	F67	2420	3500	4400	5100
203.2×12.7	F54	3060	4450	5300	6150
254.0×12.7	F69	3040	5000	6000	6850

Table 38B: CORRECTION FACTORS FOR BUSBAR DESIGN

Temperature rise		**De-rating factor**
1. Bar temperature rise 40°C over ambient 35°C		0.88
2. Bar temperature rise 30°C over ambient 35°C		0.75
Enclosure	**Cross sectional area of Busbar/Total cross sectional area of enclosure**	**De-rating factor** °
1. Outdoors	< 1% 5% 10%	0.95 0.90 0.85
2. Indoors where the enclosure itself is in a well-ventilated room	< 1% 5% 10%	0.85 0.75 0.65
3. Indoors where the enclosure itself is poorly ventilated and the room temperature is high	< 1% 5% 10%	0.65 0.60 0.50
IS:733-1967 (Wrought Aluminium and Aluminium alloys for General engineering purpose – Bars, Rods and sections) **symbol**	**Indal-Alloy**	**Rating factor** °°
EIEM°°° FIC M IIE15-WP IIE9-WP IIE30-WP NE5 M NE6 M NE20 WP	CISM 2SM B26SWP 50SWP B51SWP 54SM A56SM 65WP	1.03 1.02 0.83 0.94 0.92 0.77 0.72 0.88
ALTITUDE		
Where busbar intended for use at high altitudes is tested near sea level, the limits of temperature rise should be reduced by 1% for each 300 m in excess of 1000 m above sea level at which the busbar is to be installed.		

° These factors apply only to non-magnetic enclosures in which negligible heat is induced by the current in the busbar.

°° To be applied to the ratings for Indal D50S WP in Table 38A, the factors are approximate.

°°° Electrical-purity aluminium, material EIEM, IS: 5082-1969 (Wrought Aluminium and Aluminium alloys for electrical purpose–Bars and sections).

Table 39: DE-RATING FACTORS FOR PROXIMITY-EFFECT: THREE-PHASE CHANNEL INSTALLATION TWO CHANNELS PER IN BOX FORM

Phase spacing	De-rating factor
3-channel width	0.82
4-channel width	0.89
5-channel width	0.95
6-channel width	0.99

Table 40: CLEARANCES FOR OPEN AND ENCLOSED INDOOR AIR-INSULATED BUSBAR AND CONNECTIONS (ELECTRICALLY NON-EXPOSED)

Rated voltage kV (Up to and including)	Minimum clearance to earth in air (mm)		Minimum clearance between phases in air (mm)	
	Open	Enclosed	Open	Enclosed
0.415	19	16	25	19
0.6	25	19	31	19
3.3	51	51	51	51
6.6	64	64	89	89
11	76	76	127	127
15	102	102	165	165
22	140	140	241	241
33	222	222	356	356

Table 41: CLEARANCES FOR OPEN OUTDOOR BUSBAR AND BUSBAR CONNECTIONS NOT EXCEEDING 15 kV FOR ELECTRICALLY EXPOSED OR NON-EXPOSED INSTALLATIONS

Rated voltage (kV)	Minimum clearance to earth in air (mm)	Minimum clearance in air between phases (mm)
Not exceeding		
6.6	140	178
11	178	229
15	216	267

- The clearances quoted in tables should be regarded as the minimum values which should be used in any circumstances.
- Busbar for use in systems with special requirements or conditions requires special consideration and reference should always be made to the details included in B.S. 159: 1957 (Busbars and Busbar Connections), from which the values have been taken.

Table 42: PROTECTIVE MEASURES ON TRANSFORMERS (11 kV/433 V)

Rating	Method of installation	Primary control and protection	Secondary control and protection
Up to and including 400 kVA	Outdoor	Air break switch, lightning arrestor, D.O. Fuse	MCCB having isolation duty and E/F release and standby low set E/F relay using CT at neutral earthing conductor or ACB with proper current setting facility with 3 O/C and 1E/F release.
Up to and including 400 kVA	Indoor	11 kV, AB switch-fuse unit or VCB/GCB with 2 O/C and1E/F relays preferably with DC trip	MCCB having isolation duty and E/F release and standby low set E/F relay using CT at neutral earthing conductor or ACB with proper current setting facility with 3 O/C and 1E/F release.
Above 400 kVA and up to and including 630 kVA	Outdoor	AB switch, lightning arrestor, D.O. Fuse	ACB drawout type with 3 O/C & 1 E/F release and standby low set E/F relay using CT at neutral earthing conductor
Above 400 kVA and up to and including 630 kVA	Indoor	11 kV, ABSF unit or VCB/GCB with 2 O/C and 1E/F relays with DC trip	ACB draw out type with 3 O/C & 1 E/F release and standby low set E/F relay using CT at neutral earthing conductor
Above 630 kVA and below 1000 kVA	Indoor	11 kV, ABSF unit or VCB/GCB with 2 O/C and 1E/F relays preferably with DC trip	ACB drawout type with 3 O/C & 1 E/F release and standby low set E/F relay using CT at neutral earthing conductor
Above 1000 kVA and up to and including 2000 kVA	Indoor	VCB/GCB with 2 O/C relays with high set elements, one instantaneous E/F relay, Buchholz relay, oil and winding temperature high alarm and trip. For dry type transformers, Buchholz relay, oil temperature relay and trip are irrelevant.	ACB drawout type with 3 O/C & 1 E/F release, 2 O/C and 1 E/F relay, standby low set E/F protection using CT at neutral earthing conductor, REF protection relay with primary trip. If breaker is having microprocessor based programmable releases having different time settings, separate O/C and E/F relay need not be provided. But low setting E/F relay and REF relay are to be provided.

Table 43: INRUSH TIME CONSTANTS FOR OIL FILLED TRANSFORMERS

Transformer rating kVA S_{tr}	$k_i = \frac{i_{p_{inrush}}}{I_{1_{tr}}}$	τ_{inrush}
50	15	0.10
100	14	0.15
160	12	0.2
250	12	0.22
400	12	0.25
630	11	0.30
1000	10	0.35
1600	9	0.40
2000	8	0.45

Table 44: INRUSH TIME CONSTANTS FOR CAST RESIN TRANSFORMERS

Transformer rating kVA S_{tr}	$k_i = \frac{i_{p_{inrush}}}{I_{1_{tr}}}$	τ_{inrush}
200	10.5	0.15
250	10.5	0.18
315	10	0.20
400-500	10	0.25
630	10	0.26
800-1000	10	0.30
1250	10	0.35
1600	10	0.40
2000	9.5	0.40

(Source: ABB Ltd.)

where

S_{tr} = rated capacity of the transformer–kVA

$i_{p_{inrush}}$ = inrush current in primary winding–A

$I_{1_{tr}}$ = primary rated current–A

Table 45: SHORT CIRCUIT CURRENTS

The Bureau of Indian Standard specification IS-13234-92 and IEC standard 60909-1988 gives very useful information for the calculation of the peak current.

Initial symmetrical short circuit current $= I_k'' = \frac{cU_n}{\sqrt{3} \times Z_k}$ A

And, peak short circuit current $= i_p = k\sqrt{2} \times I_k''$

where,

U_n = Nominal line to line system voltage
Z_k = System impedance up to the fault point 'k'
c = Voltage factor
= 1 for low voltage (240/415)
= 1.05 for other low voltages
= 1.1 for voltages above 1 kV up to 230 kV

$$k = 1.02 + 0.98e^{\left(\frac{-3R_k}{X_k}\right)}$$

The value of 'k' can also be evaluated using the figure shown below.

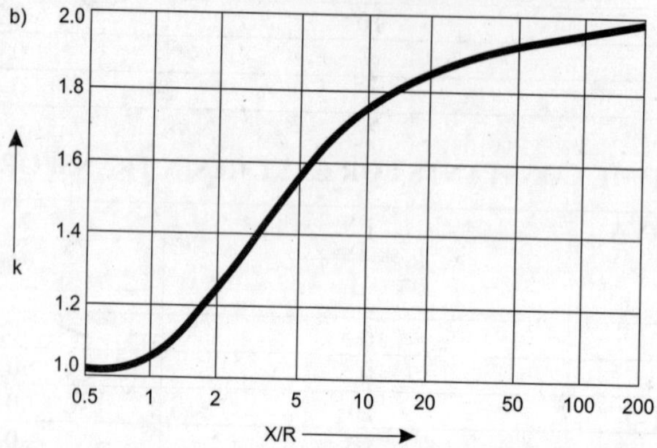

Graph showing the relation between $\frac{X}{R}$ and k

The breaking current or the rms value of the alternating current flowing through the first phase at the instant of contact separation, during the opening of the circuit breaker in the event of a short circuit is given by

$$I_{cu} = \mu I_k''$$

where, the value of μ depends on the minimum operating delay of the circuit breaker and the ratio of $\frac{I_k''}{I_n}$. The value of μ can be obtained from the figure shown below.

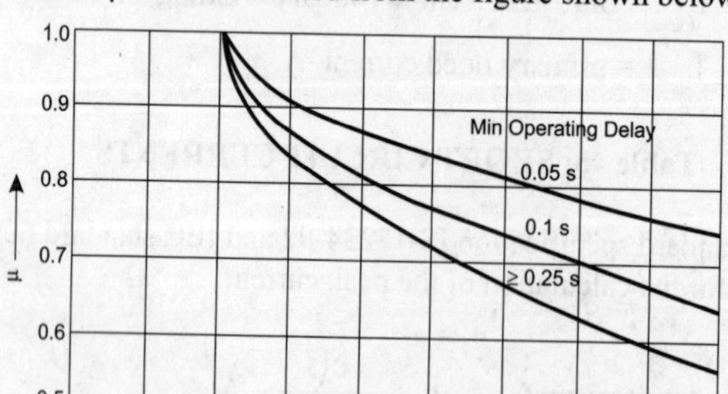

Factor μ for determining the breaking current (IS-13234)

The initial symmetrical short circuit current '$I_k^{"}$' is the current that is to be considered while selecting the size of the cable, breaking capacity of the circuit breaker and designing the earthing system. The peak short circuit current 'i_p' is to be used in deciding the making capacity of the breakers as well as the dynamic forces due to the short circuit currents on the busbars.

Table 45A: INSTANTANEOUS AND FIRST HALF CYCLE FACTORS (IEC 60947-2)

$\frac{X}{R}$ ratios	Instantaneous peak factor	Half cycle factor	Time to peak t_p (ms)
0	1.4142	1.00	5.00
0.5	1.4250	1.00	6.48
1.0	1.5122	1.002	7.32
2.0	1.7560	1.042	8.16
3.0	1.9495	1.115	8.52
4.0	2.0892	1.191	8.88
5	2.1924	1.263	9.00
6	2.2708	1.304	9.12
7	2.3323	1.347	9.24
8	2.3817	1.381	9.36
9	2.4222	1.412	9.36
10	2.4561	1.438	9.48
20	2.6250	1.570	9.72
30	2.6890	1.618	9.84
40	2.7224	1.643	9.84
50	2.7427	1.697	9.84
100	2.7848	1.697	9.96
∞	2.8284	1.732	10.00

Table 46: FAULT LEVEL COMPUTATION AND EARTHING DESIGN

Basic equations:

1 Source

Per unit source impedance $$Z_Q = \frac{\text{Base MVA}}{\text{Short circuit MVA}}$$

Source impedance in ohms $$Z_Q = \frac{cU_{nQ}^2}{S_{kQ}^{"}}$$

where,

c = Voltage factor
= 1 for 230/415 V
= 1.05 for all other low voltages

= 1.1 for voltages from 1 kV to 230 kV

U_{nQ} = Nominal source voltage-line to line (rms) at Q

S''_{kQ} = Initial symmetric short circuit apparent power at Q

2 Transmission line

Per unit line impedance $Z_{l(pu)}$ $= \dfrac{\text{Line impedance}}{\text{Base impedance}}$

Base impedance Z_b $= \dfrac{\text{Base kV}^2}{\text{Base MVA}}$

Line inductance (sym. spacing) L $= \left(0.5 + \ln\dfrac{d}{r}\right)10^{-7}$ H/m

where,
d = spacing between conductors in metres and
r = radius of conductor in metres.

Inductive reactance of the overhead line $= 0.0628\left(\dfrac{0.25}{n} + \ln\dfrac{d}{r}\right)$ Ω/km

where,
n = number of conductors per phase
d = spacing of overhead line in metres

d = mean radius of the conductor $= \dfrac{1.14}{1000}\sqrt{\dfrac{q_n}{\pi}}$ metres

q_n = area of cross-section of the conductor in mm^2

3 Transformer

S_{rT} = Rated apparent power of transformers–VA

U_{rT} = Rated voltage of transformer on HV or MV side–V

u_{kr} = Percentage impedance (impedance voltage) of the transformer

Z_T = Transformer impedance– $\Omega = \dfrac{u_{kr}}{100}\dfrac{U_{rT}^2}{S_{rT}}$

P_{krT} = Full load loss of transformer–W

R_T = Resistance of the transformer $= \dfrac{P_{krT}}{3I_{rT}^2}$

X_T = Reactance of the transformer $= \sqrt{\left(Z_T^2 - R_T^2\right)}$ Ω

I_{rT} = Full load current of the transformer on HV or MV side-A

4 Soil

Soil resistivity ρ = 2πDR ohm-m

where,
D = Spacing of the spikes in metre and
R = Reading of the earth tester in ohms.

5 Earth electrodes

Permissible current density of the earth electrode $I_d = \dfrac{7.57 \times 10^3}{\sqrt{\rho t}}$ A/m²

where,

t = Duration of fault in seconds (3 sec)

Earth electrode resistance values for different electrodes:

Plate electrode $R = \dfrac{\rho}{4}\sqrt{\dfrac{\pi}{A}}$ ohm

Pipe electrode $R = \dfrac{100\rho}{2\pi l}\ln\dfrac{2l}{d}$ ohm

Strip or conductor electrode $R = \dfrac{100\rho}{2\pi l}\ln\dfrac{4l}{t}$ ohm

where,

A = area of plate electrode (both sides) –m²
l = length of the electrode –m
d = diameter of the electrode –m
t = width of the strip or twice the diameter of the circular conductor-m

Data required:

1 Supply voltage of the system-U_n
2 Fault level on the nearest substation- S''_{kQ}
3 Length of the transmission line from the substation to the consumers transformer-l
4 Spacing of the conductors of the O/H transmission line-d
5 Radius of the conductor of the O/H line-r (Cross-section of the conductor-mm²-q_n)
6 Resistance of the line-R_l
7 kVA capacity of the consumer transformer-S_{rT}
8 Percentage impedance (impedance voltage) of the consumer transformer-u_{kT}
9 Full load loss of the transformer – P_{krT}
10 Soil resistivity of the transformer location-ρ

Design earth electrodes:

The receiving end transformer station is fed through 11 kV OH line having three line conductors. Any earth fault, i.e., line to ground fault current will have to return to the sending end substation through the earthing system and the mass of the earth to the neutral point of the sending end substation transformer neutral. Hence, all the ground fault current on the HT side of the transformer station shall dissipate to the mass of the earth through the designed earthing system. The total symmetrical fault current on the 11 kV side is taken as the ground fault current also. Therefore, the HT earthing system is designed to the calculated value of fault current on the HT side.

The number of earth electrodes can be calculated based on the fault level on the HT side of the consumer transformer.

Permissible current density of an electrode $= \dfrac{7.57 \times 10^3}{\sqrt{\rho t}}$

$= I_d$

where t = duration of fault (normally 3s)

Area of earth plates required $= \dfrac{I_A''}{I_d} \times 1000\ m^2$

(where, I_A'' = initial symmetrical rms fault current on the HV side of the transformer)

$= A_E$

Standard size of plate earth electrode $= 1.2 \times 1.2 \times 0.012$ m

Hence, the area of a plate electrodes $= 2\,(1.2 \times 1.2)\ m^2$
$= 2.88\ m^2$

Hence, number of earth electrodes required $= \dfrac{A_E}{2.88}$

(Rounded to the next integer)

Fault current dissipation on secondary side (LT side) of the transformer station

Always there is an earth grid formed in the industry, interlinking plants and the substation. All the equipment bodies are connected to the earth grid using duplicate earthing conductors and the earth grid in turn is connected to the neutral of the transformer. A few anchoring earth electrodes are provided as equipotential bonding with the mass of the earth. Since the earth grid is connected to transformer neutral, any earth fault current will find an easy low resistance path through the grid to the neutral. Thus, major portion of the LT fault current will return to the transformer neutral through the earth grid and not through the earth electrodes. As a standard, a minimum of **4 earth electrodes** will be provided on the LT side as equipotential bonding with the mass of the earth.

Determination of earthing conductor size

(Assume that copper is used as earthing conductor)

Permissible current density of copper $= 118\ A/mm^2$

Area of earth conductor on H.T side $= \dfrac{I_A''}{118}\ mm^2$

(Choose the nearest standard area of cross section)

Area of earth conductor on the MV side $= \dfrac{I_B''}{118}\ mm^2$

(where I_B'' = initial symmetrical rms fault current on the MV side of the transformer)

(Choose the nearest standard area of cross section)

Table 47: PROPERTIES OF ACSR CONDUCTORS

Code name	No. of aluminium strands	Dia in mm	No. of steel strands	Dia. in mm	Dia. of cable in mm	Gross area of aluminium mm^2	Gross area of cable mm^2	Ultimate strength kg	Resistance at 20°C – ohm/km	Current for 30° rise above ambient–A	Modulus of elasticity $kg/cm^2 \times 10^6$
Squirrel	6	2.11	1	2.11	6.33	20.71	24.44	770	1.37	70	0.81
Easel	6	2.59	1	2.59	7.77	31.63	36.90	1135	0.91	100	0.81
Rabbit	6	3.35	1	3.35	10.15	52.95	61.78	1860	0.54	148	0.81
Mink	6	3.66	1	3.66	10.98	63.06	73.57	2205	0.45	167	0.81
Dog	6	4.72	7	1.57	14.15	105.20	118.80	3310	0.22	254	0.74
Tiger	30	2.36	7	2.36	16.52	131.50	162.10	5790	0.22	296	0.79
Wolf	30	2.59	7	2.59	18.13	158.10	195.00	6875	0.18	343	0.79
Lynx	30	2.79	7	2.79	19.53	183.90	226.80	7945	0.16	385	0.79
Panther	30	3.00	7	3.00	21.00	211.70	261.20	9095	0.14	427	0.79
Kundah	42	3.49	7	1.94	26.76	402.90	423.57	9550	0.07	726	0.65
Finch	54	3.64	19	2.18	32.83	564.00	635.50	18235	0.05	1000	-

Table 48: PROPERTIES OF ALL ALUMINIUM CONDUCTORS

Code name	Strands & wire diameter in mm	Diameter of complete cable mm	Ultimate strength kg.	Resistance at 20°C ohm/km	Weight kg/km	Current rating per 30°C temp rise A	Aluminium area mm^2
Rose	7/1.96	5.9	415	1.351	57.7	100	21.2
Gnat	7/2.21	6.6	490	1.064	73.4	118	26.8
Iris	7/2.47	7.4	635	0.850	91.8	131	33.6
Lady bird	7/2.79	8.4	740	0.665	117.3	159	42.9
Ant	7/3.10	9.3	890	0.541	144.3	182	52.8
Fly	7/3.40	10.2	1,050	0.449	174.0	207	63.7
Blue Bottle	7/3.66	11.0	1,195	0.388	201.0	225	73.6
Ear wing	7/3.66	11.3	1,275	0.263	215.3	238	78.8
Grass Hopper	7/3.91	11.7	1,355	0.339	229.9	250	84.1
May bug	37/4.09	28.6	7,289	0.0598	1343.0	764	473.6
Scorpion	37/4.27	29.9	7,878	0.0549	1464.0	805	518.4

Table 49: REGULATION CONSTANTS FOR 11 kV OVERHEAD LINES

REGULATION CONSTANT (% REGULATION) PER 100 kVA km

Phase to phase voltage : 11 kV
Conductor spacing (equivalent) : 1500 mm

Conductor	6/1/3.66	6/1/3.35	6/1/2.11	6/1/2.59
Unity p.f	0.0482	0.0482	0.1219	0.0808
0.9 p.f	0.0499	0.0571	0.1244	0.0869
0.8 p.f	0.5091	0.0574	0.1178	0.0841
0.7 p.f	0.0504	0.0561	0.1094	0.0798

REGULATION CONSTANT (% REGULATION) PER 100 kVA km

Phase to phase voltage : 11 kV
Conductor spacing (equivalent) : 912 mm

Conductor	6/1/3.66	6/1/3.35	6/1/2.11	6/1/2.59
Unity p.f	0.04052	0.0482	0.1219	0.0808
0.9 p.f	0.0488	0.0559	0.1233	0.0858
0.8 p.f	0.0494	0.0558	0.1162	0.0827
0.7 p.f	0.0485	0.0543	0.1076	0.0780

Table 50: REGULATION CONSTANTS FOR 415 V OVERHEAD LINES

REGULATION CONSTANT (% REGULATION) PER 100 kVA km

Phase to phase voltage : 415 V
Conductor spacing (equivalent) : 438 mm

Conductor	7/3.91	7/3.66	7/3.4	7/3.1	7/2.79	7/2.21
Unity p.f	0.2129	0.2436	0.2813	0.3393	0.4172	0.6672
0.9 p.f	0.2654	0.2940	0.3291	0.3828	0.4547	0.6833
0.8 p.f	0.2718	0.2978	0.3296	0.3780	0.4427	0.6478
0.7 p.f	0.2699	0.2931	0.3214	0.3644	0.4217	0.6027

REGULATION CONSTANT (% REGULATION) PER 100 kVA km

Phase to phase voltage : 415 V
Conductor spacing (equivalent) : 483 mm

Conductor	7/3.91	7/3.66	7/3.4	7/3.1	7/2.79	7/2.21
Unity p.f	0.2129	0.2436	0.2813	0.3393	0.4172	0.6672
0.9 p.f	0.2669	0.2956	0.3309	0.3844	0.4562	0.6847
0.8 p.f	0.2740	0.2999	0.3314	0.3802	0.4499	0.6497
0.7 p.f	0.2122	0.2956	0.3239	0.3669	0.4240	0.6051

Regulation : Allowable values are H.T. +6% and –9%, L.T. +/-6% only.

Table 51: CABLE SELECTION CRITERIA

The following points shall be taken into consideration before selecting the cable for a particular application:

- The power and voltage rating for which the cables are being used
- Conditions of installation at the site
- Current carrying capacity of the cables
- Voltage drop in the cables
- Short circuit capacity of the cables
- Availability of the selected size of the cable

Table 52: VOLTAGE DROPS OF CABLES (in mV per ampere per meter)

Aluminium cross-section (mm^2)	Single-phase circuits			Three-phase circuits		
	Motor loads		Lighting (PF = 1)	Motor loads		Lighting (PF = 1)
	Normal running (PF = 0.8)	Starting (PF = 0.35)		Normal running (PF = 0.8)	Starting (PF = 0.35)	
6	9.1	4.1	11.2	8	3.6	9.5
10	6.1	2.9	7.5	5.3	2.5	6.2
16	3.7	1.7	4.5	3.2	1.5	3.6
25	2.36	1.15	2.8	2.05	1.0	2.4
35	1.5	0.75	1.8	1.3	0.65	1.5
50	1.15	0.6	1.29	1.0	0.52	1.1
70	0.86	0.47	0.95	0.75	0.41	0.77
120	0.64	0.37	0.64	0.56	0.32	0.55
150	0.48	0.30	0.47	0.42	0.26	0.40
185	0.39	0.26	0.37	0.34	0.23	0.31
240	0.33	0.24	0.30	0.29	0.21	0.27
300	0.29	0.22	0.24	0.25	0.19	0.20
400	0.24	0.20	0.19	0.21	0.17	0.16

Table 53: SHORT CIRCUIT RATING OF CABLES (FOR ONE SECOND DURATION) (kA)

Area of conductor (mm^2)	Aluminium		Copper	
	PVC	HRPVC	PVC	HRPVC
1.5	0.1126	0.097	0.169	0.153
2.5	0.1878	0.162	0.283	0.255
4	0.3004	0.259	0.452	0.406
6	0.4507	0.389	0.679	0.612
10	0.7512	0.648	1.132	1.020
16	1.2000	1.087	1.810	1.632
25	1.8700	1.620	2.830	2.550
35	2.6200	2.270	3.960	3.570
50	3.7500	3.243	5.660	5.100
70	5.2500	4.540	7.920	7.140
95	7.1300	6.150	10.750	9.690
120	9.0000	7.784	13.580	13.340
150	11.260	9.730	16.980	15.300
185	13.890	12.00	20.940	18.870
240	18.020	15.56	27.160	24.480
300	22.530	19.46	33.960	30.600
400	30.040	25.94	45.280	40.800
500	37.560	32.43	56.600	51.000
630	47.320	40.86	71.310	64.250

PROPERTIES OF CABLES (Source: Finolex Cables)

Table 54: 1100 V GRADE SINGLE CORE, ALUMINIUM CONDUCTOR, XLPE INSULATED, UNARMOURED/ ARMOURED, PVC JACKETED CABLES

Nom. cross sectional area	Unarmoured			Armoured				Max. DC conductor resistance at 20ºC	A.C. Current rating			
									In Air		In Ground	
	Nom. thickness of XLPE insulation	Approx. overall Dia.	Approx. overall weight	Nom. thickness of insulation	Nom. Dia of Al. wire for armour	Approx. overall Dia	Approx. weight		2 Cables	3 Cables	2 Cables	3 Cables
mm^2	mm	mm	kg/km	mm	mm	mm	kg/km	Ohm/km	Amps	Amps	Amps	Amps
10	0.7	9.7	106	1	1.4	12.1	176	3.08	60	53	69	59
16	0.7	10.7	134	1	1.4	13.2	211	1.91	82	73	90	76
25	0.9	12.3	185	1.2	1.4	14.8	276	1.2	108	99	116	97
35	0.9	13.4	223	1.2	1.4	15.9	321	0.868	136	122	139	116
50	1	15	277	1.3	1.4	17.5	387	0.641	163	149	162	139
70	1.1	17.2	363	1.4	1.4	19.7	487	0.443	208	190	199	171
95	1.1	18.7	444	1.4	1.6	21.9	617	0.32	258	235	241	204
120	1.2	20.5	538	1.5	1.6	23.6	726	0.253	303	276	273	231
150	1.4	22.7	666	1.7	1.6	25.5	850	0.206	348	321	305	259
185	1.6	24.9	797	1.9	1.6	27.7	999	0.164	407	371	347	292
240	1.7	27.3	985	2	1.6	30.1	1208	0.125	488	447	407	342
300	1.8	30.1	7786	2.1	1.6	33.2	1452	0.1	569	515	458	384
400	2	33.8	1511	2.4	2	37.5	1869	0.0778	669	606	518	440
500	2.2	37.2	1867	2.6	2	40.9	2261	0.0605	786	705	592	500
630	2.4	40.7	2321	2.8	2	44.7	2789	0.0469	922	823	666	565
800	2.6	46.4	2938	3.1	2	50.5	3473	0.0367	1067	949	750	629
1000	2.8	50.7	3638	3.3	2.5	55.7	4364	0.0291	1220	1076	833	704

Table 55: 1100 V GRADE TWO CORE, ALUMINIUM CONDUCTOR, XLPE INSULATED, UNARMOURED/ ARMOURED, PVC JACKETED CABLES

Nom. cross sectional area	Nom. thickness of XLPE insulation	Unarmoured		Armoured				Max. DC conductor resistance at 20°C	A.C. current rating	
		Approx. overall Dia.	Approx. overall weight	Nom. steel armour size mm		Approx. overall Dia.	Approx. weight		In air	In ground
mm^2	mm	mm	kg/km	Round wire	Flat strip	mm	kg/km	Ohm/km	Amps	Amps
10	0.7	17.3	132	1.4		18.1	530	3.08	67	71
16	0.7	19	147	1.4		20.4	643	1.91	88	91
25	0.9	19.6	166		4 × 0.8	19.3	582	1.2	117	120
35	0.9	21.2	181		4 × 0.8	20.9	680	0.868	145	143
50	1	23.7	204		4 × 0.8	23.4	821	0.641	176	167
70	1.1	27.1	236		4 × 0.8	27.1	1054	0.443	221	204
95	1.1	30	286		4 × 0.8	29.4	1271	0.32	271	245
120	1.2	32.7	314		4 × 0.8	32.1	1495	0.253	315	278
150	1.4	34.5	333		4 × 0.8	34.3	1769	0.206	362	315
185	1.6	38.3	401		4 × 0.8	37.9	2101	0.164	420	356
240	1.7	42.2	476		4 × 0.8	41.4	2534	0.125	497	407
300	1.8	46.7	566		4 × 0.8	46.3	3079	0.1	578	463
400	2	52.3	678		4 × 0.8	51.8	3834	0.0778	678	528

Table 56: 1100 V GRADE THREE CORE, ALUMINIUM CONDUCTOR, XLPE INSULATED, UNARMOURED/ ARMOURED, PVC JACKETED CABLES

Nom. cross-sectional area	Nom. thickness of XLPE insulation	Unarmoured		Armoured				Max. DC conductor resistance at 20°C	A.C. current rating	
		Approx. overall Dia	Approx. overall weight	Nom. steel armour size mm		Approx. overall Dia	Approx. weight		In Air	In Ground
mm²	mm	mm	kg/km	Round wire	Flat strip	mm	kg/km	Ohm/km	Amps	Amps
10	0.7	18.2	294	1.4		18.9	659	3.08	53	57
16	0.7	18.4	372		4 × 0.8	18.2	540	1.91	70	78
25	0.9	21.5	528		4 × 0.8	21.2	727	1.2	95	97
35	0.9	23.5	640		4 × 0.8	23.2	863	0.868	117	116
50	1	26.7	819		4 × 0.8	26	1053	0.641	140	134
70	1.1	31.1	1101		4 × 0.8	30.5	1386	0.443	176	167
95	1.1	33.6	1359		4 × 0.8	33	1671	0.32	221	199
120	1.2	36.8	1633		4 × 0.8	36.2	1982	0.253	258	227
150	1.4	39.3	1995		4 × 0.8	38.9	2386	0.206	294	255
185	1.6	43.5	2428		4 × 0.8	43	2854	0.164	339	287
240	1.7	48.1	3043		4 × 0.8	47.7	3528	0.125	402	333
300	1.8	53.8	3760		4 × 0.8	52.8	4232	0.1	461	375
400	2	59.8	4723		4 × 0.8	59.3	5318	0.0778	542	426

Table 57: 1100 V GRADE 3.5 CORE, ALUMINIUM CONDUCTOR, XLPE INSULATED, UNARMOURED/ARMOURED, PVC JACKETED CABLES

Nom. cross-sectional area	Nom. thickness of XLPE insulation	Unarmoured		Armoured			Max. DC conductor resistance at 20°C	A.C. current rating	
		Approx. overall Dia	Approx. overall weight	Nom. steel armour size	Approx. overall Dia	Approx. weight		In Air	In Ground
mm^2	mm	mm	kg/km	mm	mm	kg/km	Ohm/km	Amps	Amps
25	0.9	23.1	604	4 × 0.8	22.8	821	1.2	95	97
35	0.9	25.2	719	4 × 0.8	24.9	961	0.868	117	116
50	1	28.8	940	4 × 0.8	28.1	1195	0.641	140	134
70	1.1	33.6	1258	4 × 0.8	33	1569	0.443	176	167
95	1.1	36.4	1560	4 × 0.8	35.8	1903	0.32	221	199
120	1.2	39.2	1896	4 × 0.8	39	2303	0.253	258	227
150	1.4	43.3	2284	4 × 0.8	42.9	2720	0.206	294	255
185	1.6	48	2800	4 × 0.8	47.5	3276	0.164	339	287
240	1.7	53.1	3506	4 × 0.8	52.7	4048	0.125	402	333
300	1.8	59.3	4343	4 × 0.8	58.4	4872	0.1	461	375
400	2	66.4	5492	4 × 0.8	65.6	6101	0.0778	542	426

Table 58: 1100 V GRADE FOUR CORE, ALUMINIUM CONDUCTOR, XLPE INSULATED, UNARMOURED/ARMOURED, PVC JACKETED CABLES

Nom. Cross-sectional Area	Nom. thickness of XLPE insulation	Unarmoured		Armoured				Max. DC Conductor Resistance at 20°C	A.C. current rating	
		Approx. overall Dia	Approx. overall weight	Nom. steel armour size mm		Approx. overall Dia	Approx. weight		In Air	In Ground
mm^2	mm	mm	kg/km	Round wire	Flat strip	mm	kg/km	Ohm/km	Amps	Amps
10	0.7	19.7	367	1.4		21.1	713	3.08	53	57
16	0.7	20	449		4 × 0.8	20.2	653	1.91	70	78
25	0.9	24	652		4 × 0.8	23.7	882	1.2	95	97
35	0.9	26.2	798		4 × 0.8	26	1055	0.868	117	116
50	1	29.6	1002		4 × 0.8	29.7	1317	0.641	140	134
70	1.1	35.1	1387		4 × 0.8	34.6	1717	0.443	176	167
95	1.1	38.1	1726		4 × 0.8	37.6	2091	0.32	221	199
120	1.2	42.3	2123		4 × 0.8	42	2550	0.253	258	227
150	1.4	45.2	2597		4 × 0.8	44.8	3042	0.206	294	255
185	1.6	50.6	3203		4 × 0.8	49.7	3647	0.164	339	287
240	1.7	55.8	4016		4 × 0.8	55	4523	0.125	402	333
300	1.8	62	4900		4 × 0.8	61.4	5482	0.1	461	375
400	2	69.4	6223		4 × 0.8	68.6	6846	0.0778	542	426

Table 59: 1100 V GRADE, ONE CORE, ALUMINIUM CONDUCTOR, PVC INSULATED, PVC SHEATHED, UNARMOURED /ARMOURED CABLES

Nom. cross-sectional area	Unarmoured			Nom. thickness of PVC	Armoured			Max. DC conductor resistance at 20°C	A.C. Current rating							
									In Air				In Ground			
	Nom. thickness of PVC Insulation	Approx. overall Dia	Approx. overall weight	Insulation	Nom. Dia of Al. wire for armour	Approx. overall Dia	Approx. weight		GPPVC Insulation		HRPVC Insulation		GPPVC Insulation		HRPVC Insulation	
									2 Cables	3 Cables	2 Cables	3 Cables	2 Cables	3 Cables	2 Cables	3 Cables
Sq.mm	mm	mm	kg/km	mm	mm	mm	kg/km	Ohm/km	Amps	Amps	Amps	Amps	Amps	Amps	Amps	Amps
2.5	0.9	8	70	1.2	1.4	10.2	135	12.1	25	21	30	26	28	24	33	28
4	1	8.7	81	1.3	1.4	10.9	152	7.41	32	27	38	33	36	31	41	35
6	1	9.2	100	1.3	1.4	11.5	170	4.61	41	35	50	42	44	39	52	46
10	1	10.2	120	1.3	1.4	12.9	200	3.08	56	47	69	57	59	51	69	60
16	1	11.5	150	1.3	1.4	13.8	240	1.91	72	64	88	78	75	66	88	77
25	1.2	13.3	210	1.5	1.4	15.6	312	1.2	99	84	121	103	97	86	112	101
35	1.2	14.4	250	1.5	1.4	16.6	365	0.868	120	105	146	129	120	100	140	117
50	1.4	15.8	320	1.7	1.4	18	445	0.641	150	130	184	158	145	120	170	140
70	1.4	18.1	415	1.7	1.4	20.7	560	0.443	185	155	226	190	170	140	199	163
95	1.6	19.9	520	1.9	1.6	22.9	715	0.32	215	190	264	233	205	175	240	205
120	1.6	22.2	630	1.9	1.6	24.8	825	0.253	240	220	292	270	230	195	268	228
150	1.8	22.6	750	2.1	1.6	26.8	960	0.206	270	250	330	305	265	220	310	258
185	2	26	900	2.3	1.6	28.6	1125	0.164	305	290	374	356	300	240	351	278
240	2.2	29	1125	2.5	1.6	31.9	1390	0.125	350	335	430	410	335	270	392	316
300	2.4	31.7	1350	2.7	2	34.6	1650	0.1	395	380	484	464	370	295	434	345
400	2.6	35.7	1720	3	2	39.2	2115	0.0778	455	435	557	534	410	325	480	378
500	3	39	2140	3.4	2	42.9	2635	0.0605	490	480	600	589	435	345	510	403
630	3.4	43.7	2710	3.9	2	47.7	3250	0.0469	560	550	680	674	485	390	565	456
800	3.4	47.2	3300	3.9	2	51.2	3915	0.0367	640	630	794	785	530	440	620	515
1000	3.4	51.5	4050	3.9	2.5	56.4	4820	0.0291	740	720	895	800	580	490	680	570

Table 60: 1100 V GRADE, TWO CORE, ALUMINIUM CONDUCTOR, PVC INSULATED, PVC SHEATHED, UNARMOURED/ARMOURED CABLES

Nom. cross-sectional area	Nom. thickness of PVC insulation	Unarmoured		Armoured			Max. DC conductor resistance at 20°C	A.C. current rating			
								In Air		In Ground	
		Approx. overall Dia	Approx. overall weight	Nom. steel armour size	Approx. overall Dia	Approx. weight		GPPVC Insulation	HRPVC Insulation	GPPVC Insulation	HRPVC Insulation
mm^2	mm	mm	kg/km	mm W F	mm	kg/km	Ohm/km	Amps	Amps	Amps	Amps
2.5	0.9	13.8	165	1.4	15.5	440	12.1	21	26	25	29
4	1	15.5	200	1.4	15.8	510	7.41	27	33	32	36
6	1	16.5	230	1.4	17	565	4.61	35	42	40	47
10	1	18.4	275	1.4	19.2	620	3.08	47	57	55	64
16	1	20.5	364	4 × 0.8	21.2	620	1.91	59	72	70	82
25	1.2	22	480	4 × 0.8	21.2	710	1.2	78	95	90	105
35	1.2	23.5	570	4 × 0.8	22.5	825	0.868	99	100	110	128
50	1.4	26	710	4 × 0.8	25	1010	0.641	125	153	135	158
70	1.4	30.4	920	4 × 0.8	28.5	1250	0.443	150	183	160	187
95	1.6	33.5	1175	4 × 0.8	31	1575	0.32	185	225	190	222
120	1.6	37	1375	4 × 0.8	33.5	1790	0.253	210	256	210	245
150	1.8	37	1660	4 × 0.8	37.5	2065	0.206	240	294	240	280
185	2	40.5	1960	4 × 0.8	40.5	2515	0.164	275	337	275	322
240	2.2	45	2470	4 × 0.8	45	3070	0.125	325	397	320	374
300	2.4	49	3000	4 × 0.8	49.5	3705	0.1	365	445	355	416
400	2.6	56	3830	4 × 0.8	55.5	4550	0.0778	420	515	385	450

Table 61: 1100 V GRADE, THREE CORE, ALUMINIUM CONDUCTOR, PVC INSULATED, PVC SHEATHED, UNARMOURED/ARMOURED CABLES

Nom. cross-sectional area	Nom. thickness of PVC insulation	Unarmoured		Armoured			Max. DC conductor resistance at 20°C	A.C. current rating			
								In Air		In Ground	
		Approx. overall Dia	Approx. overall weight	Nom. steel armour size	Approx. overall Dia	Approx. weight		GPPVC Insulation	HRPVC Insulation	GPPVC Insulation	HRPVC Insulation
mm^2	mm	mm	kg/km	mm W F	mm	kg/km	Ohm/km	Amps	Amps	Amps	Amps
2.5	0.9	14.4	195	1.4	16.1	445	12.1	18	22	21	24
4	1	16	235	1.4	16.5	520	7.41	23	28	28	32
6	1	17.5	270	1.4	18	580	4.61	30	38	35	41
10	1	18.5	335	1.4	20	695	3.08	40	49	46	54
16	1	21.5	455	4 × 0.8	22.5	730	1.91	51	63	60	69
25	1.2	23	630	4 × 0.8	23.5	900	1.2	70	86	76	89
35	1.2	25	755	4 × 0.8	25	1050	0.868	86	105	92	108
50	1.4	27.5	960	4 × 0.8	28.5	1320	0.641	105	127	110	129
70	1.4	32	1250	4 × 0.8	32.5	1635	0.443	130	159	135	157
95	1.6	35.5	1615	4 × 0.8	36	2045	0.32	155	190	165	193
120	1.6	39	1935	4 × 0.8	39	2415	0.253	180	218	185	215
150	1.8	43	2300	4 × 0.8	43.5	2810	0.206	205	251	210	246
185	2	46.5	2780	4 × 0.8	46	3305	0.164	240	294	235	276
240	2.2	52	3510	4 × 0.8	52.5	4140	0.125	280	340	275	322
300	2.4	57.5	4330	4 × 0.8	58	5040	0.1	315	386	305	358
400	2.6	65	5450	4 × 0.8	65	6185	0.0778	375	445	335	390

Table 62: 1100 V GRADE, 3.5 CORE, ALUMINIUM CONDUCTOR, PVC INSULATED, PVC SHEATHED, UNARMOURED /ARMOURED CABLES

Nom. cross-sectional area	Nom. thickness of PVC insulation	Unarmoured		Armoured			Max. DC conductor resistance at 20ºC	A.C. current rating			
								In Air		In Ground	
		Approx. overall Dia	Approx. overall weight	Nom. steel armour size	Approx. overall Dia	Approx. weight		GPPVC Insulation	HRPVC Insulation	GPPVC Insulation	HRPVC Insulation
mm^2	mm	mm	kg/km	mm W F	mm	kg/km	Ohm/km	Amps	Amps	Amps	Amps
25	1.2	24	730	4 × 0.8	24.5	970	1.2	70	86	76	89
35	1.2	26	860	4 × 0.8	26.5	1130	0.868	86	105	92	108
50	1.4	29.5	1140	4 × 0.8	30	1440	0.641	105	127	110	129
70	1.4	34	1410	4 × 0.8	34.5	1815	0.443	130	159	135	157
95	1.6	37.5	1810	4 × 0.8	38	2250	0.32	155	190	165	193
120	1.6	41.5	2255	4 × 0.8	42	2690	0.253	180	218	185	215
150	1.8	46.5	2650	4 × 0.8	46	3160	0.206	205	251	210	246
185	2	49.5	3235	4 × 0.8	50	3790	0.164	240	294	235	276
240	2.2	56	4110	4 × 0.8	56	4700	0.125	280	340	275	322
300	2.4	62	5060	4 × 0.8	62	5665	0.1	315	386	305	358
400	2.6	69.5	5985	4 × 0.8	70	7030	0.0778	375	445	335	390

Table 63: 1100 V GRADE, FOUR CORE, ALUMINIUM CONDUCTOR, PVC INSULATED, PVC SHEATHED, UNARMOURED /ARMOURED CABLES

Nom. cross-sectional area	Nom. thickness of PVC insulation	Unarmoured		Armoured			Max. DC conductor resistance at 20°C	A.C. current rating			
								In Air		In Ground	
		Approx. overall Dia	Approx. overall weight	Nom. steel armour size	Approx. overall Dia	Approx. weight		GPPVC Insulation	HRPVC Insulation	GPPVC Insulation	HRPVC Insulation
mm^2	mm	mm	kg/km	mm W F	mm	kg/km	Ohm/km	Amps	Amps	Amps	Amps
2.5	0.9	15.5	220	1.4	17	460	12.1	18	22	21	24
4	1	17.5	275	1.4	19	54	7.41	23	28	28	32
6	1	18.5	320	1.4	20.5	615	4.61	30	38	35	41
10	1	21.5	400	4 × 0.8	22	610	3.08	40	49	46	54
16	1	24	575	4 × 0.8	24	810	1.91	51	63	60	69
25	1.2	24.5	810	4 × 0.8	25.5	1045	1.2	70	86	76	89
35	1.2	26.5	975	4 × 0.8	27.5	1240	0.868	86	105	92	108
50	1.4	30	1275	4 × 0.8	31.5	1590	0.641	105	127	110	129
70	1.4	34.5	1625	4 × 0.8	36	1995	0.443	130	159	135	157
95	1.6	38.5	2100	4 × 0.8	40	2510	0.32	155	190	165	193
120	1.6	42	2480	4 × 0.8	44	2990	0.253	180	218	185	215
150	1.8	48.5	3070	4 × 0.8	48	3545	0.206	205	251	210	246
185	2	53	3700	4 × 0.8	52.5	4255	0.164	240	294	235	276
240	2.2	59.5	4675	4 × 0.8	59	5325	0.125	280	340	275	322
300	2.4	65.5	5810	4 × 0.8	65	6430	0.1	315	386	305	358
400	2.6	73.5	7240	4 × 0.8	73	7980	0.0778	375	445	335	390

Table 64: 1100 V GRADE, 2, 3 AND 4 CORE, COPPER CONDUCTOR, PVC INSULATED, PVC SHEATHED, UNARMOURED/ARMOURED CABLES

No. of cores	Nom. cross-sectional area	Nom. thickness of PVC insulation	Unarmoured		Armoured			Max. DC conductor resistance at 20°C	A.C. current rating			
									In Air		In Ground	
			Approx. overall Dia	Approx. overall weight	Nom. steel armour size	Approx. overall Dia	Approx. weight		GPPVC Insulation	HRPVC Insulation	GPPVC Insulation	HRPVC Insulation
	mm^2	mm	mm	kg/km	mm W F	mm	kg/km	Ohm/km	Amps	Amps	Amps	Amps
2	4	1	15.5	250	1.4	16	520	4.61	35	43	41	48
2	6	1	16.5	300	1.4	17	590	3.08	45	55	50	58
2	10	1	18.5	400	1.4	19	740	1.83	60	73	70	81
2	16	1	20.5	555	4 × 0.8	21	820	1.15	78	95	90	105
3	4	1	16	310	1.4	16.5	590	4.61	30	37	36	42
3	6	1	17.5	380	1.4	18	690	3.08	39	48	45	53
3	10	1	19.5	520	1.4	20	875	1.83	52	63	60	69
3	16	1	21.5	755	4 × 0.8	22.5	1025	1.15	66	81	77	90
4	4	1	17.5	375	1.4	19	685	4.61	30	37	36	42
4	6	1	18.5	470	1.4	20.5	810	3.08	39	48	45	53
4	10	1	21.5	650	4 × 0.8	22	897	1.83	52	63	60	69
4	16	1	23	800	4 × 0.8	24.5	1250	1.15	66	81	77	90

Table 65: 1100 V GRADE,1.5 mm² SOLID COPPER CONDUCTOR, PVC INSULATED, PVC SHEATHED, UNARMOURED /ARMOURED CABLES

Nom. cross-sectional area	Nom. thickness of PVC insulation	Unarmoured		Armoured			Max. DC conductor resistance at 20°C	A.C. current rating			
								In Air		In Ground	
		Approx. overall Dia	Approx. overall weight	Nom. steel armour size	Approx. overall Dia	Approx. weight		GPPVC Insulation	HRPVC Insulation	GPPVC Insulation	HRPVC Insulation
mm^2	mm	mm	kg/km	mm W F	mm	kg/km	Ohm/km	Amps	Amps	Amps	Amps
2	0.8	12	155	1.4	14	375	12.1	20	24	23	27
3	0.8	12.5	175	1.4	14.5	415	12.1	17	21	21	24
4	0.8	13.5	205	1.4	15	465	12.1	17	21	21	24
5	0.8	14.5	245	1.4	16	525	12.1	14	17	17	19
6	0.8	15.5	265	1.4	17	585	12.1	13	16	15	18
7	0.8	15.5	275	1.4	17	595	12.1	12	15	14	17
9	0.8	18.5	350	1.4	20	765	12.1	12	14	14	16
10	0.8	18.5	375	1.4	20.5	785	12.1	11	13	13	15
12	0.8	18	425	4 ×0.8	18.5	690	12.1	10	12	12	14
14	0.8	19	475	4 × 0.8	20	770	12.1	10	12	11	13
16	0.8	20	535	4 × 0.8	21	845	12.1	9	11	11	12
19	0.8	21.5	625	4 × 0.8	21.5	930	12.1	9	10	10	12
24	0.8	24.5	770	4 × 0.8	25	1160	12.1	8	10	9	11
27	0.8	25	845	4 × 0.8	25.5	1240	12.1	8	9	9	10
30	0.8	26.5	920	4 × 0.8	27	1330	12.1	7	9	9	10
37	0.8	28	1100	4 × 0.8	28	1540	12.1	7	8	8	9
44	0.8	31	1290	4 × 0.8	31.5	1805	12.1	6	7	7	8
52	0.8	32	1525	4 × 0.8	32.5	2020	12.1	6	7	7	8
61	0.8	34.5	1750	4 × 0.8	35	2275	12.1	6	7	6	7

Table 66: 1100 V GRADE, 2.5 mm² SOLID COPPER CONDUCTOR, PVC INSULATED, PVC SHEATHED, UNARMOURED /ARMOURED CABLES

Nom. cross-sectional area	Nom. thickness of PVC insulation	Unarmoured		Armoured			Max. DC conductor resistance at 20°C	A.C. current rating			
								In Air		In Ground	
		Approx. overall Dia	Approx. overall weight	Nom. steel armour size mm	Approx. overall Dia	Approx. weight		GPPVC Insulation	HRPVC Insulation	GPPVC Insulation	HRPVC Insulation
mm^2	mm	mm	kg/km	W F	mm	kg/km	Ohm/km	Amps	Amps	Amps	Amps
2	0.9	13.5	200	1.4	15	440	7.41	27	32	32	37
3	0.9	14	230	1.4	15.5	495	7.41	24	29	27	31
4	0.9	15	275	1.4	16.5	560	7.41	24	29	27	31
5	0.9	16	330	1.4	18	640	7.41	19	23	23	27
6	0.9	17	350	1.4	19	720	7.41	18	21	21	25
7	0.9	17	375	1.4	19	740	7.41	17	20	20	23
9	0.9	21	480	4 × 0.8	20.5	795	7.41	16	19	19	22
10	0.9	21	515	4 × 0.8	20.5	830	7.41	15	18	18	20
12	0.9	21.7	605	4 × 0.8	21.5	915	7.41	14	17	17	19
14	0.9	22.7	685	4 × 0.8	22.5	1005	7.41	13	16	16	18
16	0.9	23.5	770	4 × 0.8	23.5	1110	7.41	12	15	15	17
19	0.9	24.5	880	4 × 0.8	24.5	1235	7.41	12	14	14	16
24	0.9	28	1090	4 × 0.8	29.5	1560	7.41	11	13	13	15
27	0.9	28.5	1200	4 × 0.8	30	1680	7.41	10	12	12	14
30	0.9	29.5	1310	4 × 0.8	31	1810	7.41	10	12	12	13
37	0.9	32.5	1615	4 × 0.8	33.6	2110	7.41	9	11	11	12
44	0.9	35	1900	4 × 0.8	37	2450	7.41	9	10	10	11
52	0.9	36	2190	4 × 0.8	39.5	2810	7.41	8	9	10	11
61	0.9	40	2525	4 × 0.8	41	3180	7.41	8	9	9	10

Table 67: 11/11 kV, THREE CORE ALUMINIUM CONDUCTOR, XLPE INSULATED AND PVC SHEATHED, SCREENED, FLAT STRIP ARMOURED CABLES. [A2XFY (P)]

	Appx. Weight of cable	Max DC Resistance at 20°C	Max DC Resistance at 90°C	Appx. Reactance at 50Hz	Appx. Capacitance per phase	Continuous Current rating		Short circuit rating for 1 second
						In Ground at 20°C	In air at 30°C	
mm^2	kg/km	Ω/km	Ω/km	Ω/km	μF/km	A	A	kA (rms)
35	3560	0.8680	1.110	0.136	0.159	120	146	3.29
50	4000	0.6410	0.8220	0.129	0.175	141	174	4.7
70	4510	0.4430	0.5680	0.119	0.196	172	216	6.58
95	5130	0.3200	0.4110	0.113	0.224	205	260	8.93
120	5670	0.2530	0.3250	0.110	0.241	233	282	11.3
150	6220	0.2060	0.2650	0.106	0.259	261	321	14.1
185	6970	0.1640	0.2110	0.103	0.281	294	364	17.4
240	7990	0.1250	0.1620	0.0985	0.314	340	429	22.6
300	9050	0.1000	0.1300	0.0952	0.340	384	492	28.2
400	10570	0.0778	0.1020	0.0929	0.375	439	575	37.6
500	12280	0.0605	0.0810	0.089	0.417	497	661	47

Table 68: METHODS FOR LAYING THE CABLES AND CONDUCTORS (IEC 60364-5-52)

Type of cables or conductors		Methods of cable installation						
		Clipped direct	Conduit	Cable trunking	Cable ducting	Cable trays/ ladders	On insulators	On support wire
Bare conductor		No	No	No	No	No	Yes	No
Insulated conductors		No	Yes	Yes	Yes	No	Yes	No
Sheathed cables including armoured cables	One core	Yes	Yes	Yes	Yes	Yes	-	Yes
	Multi core	Yes	Yes	Yes	Yes	Yes	-	Yes

Table 69: DIFFERENT METHODS OF INSTALLATION (REF: IEC 60364-5-52)

Item No.	Description	Reference method of installation to obtain current carrying capacity
1	Insulated conductors or single core cables in conduit in a thermally insulated wall	A_1
2	Multi core cables in conduit in a thermally insulated wall	A_2
3	Insulated conductors or single core cables in conduit on a wooden or masonry wall or spaced less than 0.3 times diameter of cables from the wall	B_1
4	Multi core cables on a wooden or masonry wall or spaced less than 0.3 times the diameter of the cable from it	B_2
5	Single core or multi core cables-fixed on, or spaced less than 0.3 times the cable diameter from the wall	C
6	Cables on non perforated trays	C
7	Multi core cables in conduit or in cable ducting in ground / Single core cables in conduit or in cable ducting in ground	D
8	Cables on perforated trays	E or F
9	Bare ir insulated conductors on insulators	G

Table 70: MAXIMUM OPERATING TEMPERATURES OF INSULATION MATERIALS (IEC: 60364-5-52)

Type of insulation	Temperature limit °C
Polyvinyl Chloride (PVC)	70 at the conductor
Cross-linked polyethylene (XLPE) and Ethylene propylene rubber (EPR)	90 at the conductor
Mineral (PVC coated or exposed to touch)	70 at the sheath
Mineral (Bare not exposed to touch and not in contact with combustible material)	105 at the sheath

DE-RATING OF CABLES

To account for the environmental and other conditions of installation, correction factors are to be introduced to determine the correct cross sectional area of the cables. If I_L is the actual load current, the corrected value of the load current to be taken into account for determining the cable cross section,

$$I_L' = \frac{I_L}{k_1 k_2 ....}$$

where,

k_1, k_2, …etc. are the correction or de-rating factors of the cables.

Table 71: DE-RATING FACTOR K_1 FOR CABLES LAID IN AIR (IEC 60364-5-52)

Ambient temperature °C	Insulation material of the cable	
	PVC	XLPE and EPR
15	1.17	1.12
20	1.12	1.08
25	1.06	1.04
35	0.94	0.96
40	0.87	0.91
45	0.79	0.87
50	0.71	0.82
55	0.61	0.76
60	0.50	0.71
65		0.65
70		0.58

Table 72: DE-RATING FACTOR K_2 FOR CABLES LAID IN GROUND (IEC 60364-5-52)

Ground temperature °C	Insulation material of the cable	
	PVC	XLPE and EPR
15	1.05	1.04
25	0.95	0.96
30	0.89	0.93
35	0.84	0.89
40	0.77	0.85
45	0.71	0.80
50	0.63	0.76

55	0.55	0.71
60	0.45	0.65
65		0.60
70		0.53

Table 73: DE-RATING FACTOR K_3 FOR THE NATURE OF THE SOIL (Source: Schneider Electric)

Nature of the soil	Correction factor k_3
Very wet soil	1.21
Wet soil	1.05
Damp soil	1.0
Very dry soil	0.86

Table 74: DE-RATING FACTOR K_4 FOR THE BUNCHING OF CABLES (IEC 60364-5-52)

Bunching arrangement	Number of circuits or multi core cables						Reference method
	1	2	3	4	6	9	
Bunched in air, on a surface, embedded or enclosed	1.0	0.8	0.70	0.65	0.57	0.5	A to F
Single layer on wall, on floor or unperforated tray	1.0	0.85	0.79	0.75	0.72	0.70	C
Single layer fixed directly under a wooden ceiling	0.95	0.81	0.72	0.68	0.64	0.61	C
Single layer on a perforated horizontal or vertical tray	1.0	0.88	0.82	0.77	0.73	0.72	E and F
Single layer on a ladder support or cleats, etc.	1.0	0.87	0.82	0.80	0.79	0.78	E and F

Table 75: DE-RATING FACTOR K_4 FOR MORE THAN ONE CIRCUIT, SINGLE CORE OR MULTI CORE CABLES LAID DIRECT IN THE GROUND (IEC 60364-5-52)

Number of circuits	Cable to cable clearance				
	Touching each other	One cable diameter	0.125 m	0.25 m	0.5 m
2	0.75	0.8	0.85	0.9	0.9
3	0.65	0.7	0.75	0.8	0.85
4	0.6	0.6	0.70	0.75	0.80
5	0.55	0.55	0.65	0.7	0.80
6	0.50	0.55	0.60	0.7	0.80

Table 76: DE-RATING FACTOR K_5 FOR ACCOUNTING 3rd HARMONIC CURRENT (IEC 60364-5-52).

Third harmonics %	Correction factor	
	Selection based on phase current	Selection based on neutral current
0-15	1.0	
15-33	0.86	
33-45		0.86
> 45		1.0

Table 77: CURRENT CARRYING CAPACITY OF BURIED CABLES (IEC 60364-5-52)

Installation method	Size mm²	No. of loaded conductors and type of insulation			
		Two PVC	Three PVC	Two XLPE	Three XLPE
D	Copper				
	1.5	22	18	26	22
	2.5	29	24	34	29
	4	38	31	44	37
	6	47	39	56	46
	10	63	52	73	61
	16	81	67	95	79
	25	104	86	121	101
	35	125	103	146	122
	50	148	122	173	144

	70	183	151	213	178
	95	216	179	252	211
	120	246	203	287	240
	150	278	230	324	271
	185	312	258	363	304
	240	361	297	419	351
	300	408	336	474	396
D	Aluminium				
	2.5	22	18.5	26	22
	4	29	24	34	29
	6	36	30	42	36
	10	48	40	56	47
	16	62	52	73	61
	25	80	66	93	78
	35	96	80	112	94
	50	113	94	132	112
	70	140	117	163	138
	95	166	138	193	164
	120	189	157	220	186
	150	213	178	249	210
	185	240	200	279	236
	240	277	230	322	272
	300	313	260	364	308

Table 78: CURRENT CARRYING CAPACITIES OF UNBURIED CABLES (IEC 60364-5-52)

Reference Method	Number of loaded conductors and type of insulation											
A1		2 PVC	3 PVC		3 XLPE	2 XLPE						
A2	3 PVC	2 PVC		3 XLPE	2 XLPE							
B1				3 PVC	2 PVC		3 XLPE					
B2			3 PVC	2 PVC		3 XLPE	2 XLPE		2 XLPE			
C					3 PVC		3 PVC	3 XLPE		2 XLPE		
E						3 PVC		2 PVC	3 XLPE		2 XLPE	
F							3 PVC		2 PVC	3 XLPE		2 XLPE
1	2	3	4	5	6	7	8	9	10	11	12	13
Copper (mm²)	Current carrying capacity ampere											
1.5	13	13.5	14.5	15.5	17	18.5	19.5	22	23	24	26	-
2.5	17.5	18	19.5	21	23	25	27	30	31	33	36	-

4	23	24	26	28	31	34	36	40	42	45	49	-
6	29	31	34	36	40	43	46	51	54	58	63	-
10	39	42	46	50	54	60	63	70	75	80	86	-
16	52	56	61	68	73	80	85	94	100	107	115	-
25	68	73	80	89	95	101	110	119	127	135	149	161
35	-	-	-	110	117	126	137	147	158	169	185	200
50	-	-	-	134	141	153	167	179	192	207	225	242
70	-	-	-	171	179	196	213	229	246	268	289	310
95	-	-	-	207	216	238	258	278	298	328	352	377
120	-	-	-	239	249	276	299	322	346	382	410	437
150	-	-	-	-	285	318	344	371	395	441	473	504
185	-	-	-	-	324	362	392	424	450	506	542	575
240	-	-	-	-	380	424	461	500	538	599	641	679
Aluminum (mm²)	**Current carrying capacity -ampere**											
2.5	13.5	14	15	16.5	18.5	19.5	21	23	24	26	28	-
4	17.5	18.5	20	22	25	26	28	31	32	35	38	-
6	23	24	26	28	32	33	36	39	42	45	49	-
10	31	32	36	39	44	46	49	54	58	62	67	-
16	41	43	48	53	58	61	66	73	77	84	91	-
25	53	57	63	70	73	78	83	90	97	101	108	121
35	-	-	-	86	90	96	103	112	120	126	135	150
50	-	-	-	104	110	117	125	136	146	154	164	184
70	-	-	-	133	140	150	160	174	187	198	211	237
95	-	-	-	161	170	183	195	211	227	241	257	289
120	-	-	-	186	197	212	226	245	263	280	300	337
150	-	-	-	-	226	245	261	283	304	324	346	389
185	-	-	-	-	256	280	298	323	347	371	397	447
240	-	-	-	-	300	330	352	382	407	439	470	530

Table 79: PROPERTIES OF SUBMERSIBLE FLAT CABLES (SOURCE: FINOLEX CABLES) 3 CORE FLAT CABLES WITH COPPER CONDUCTOR XLPE INSULATION AND HR PVC SHEATH TECHNICAL DATA

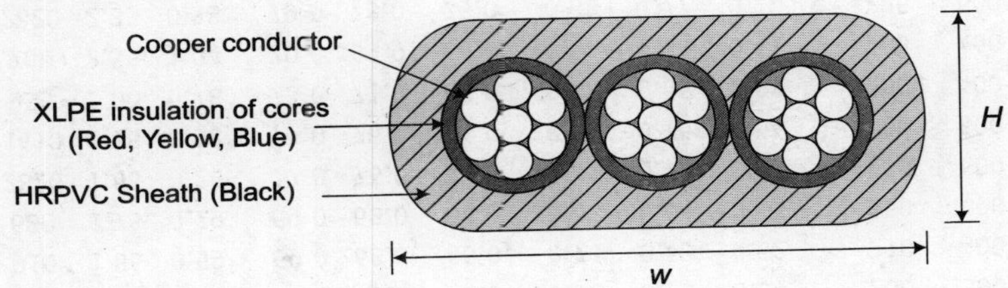

- Conductor: Flexible annealed electrolytic grade bare copper
- Insulation: Cross linked Polyethylene (XLPE)
- Sheath: Heat Resistant PVC Type ST2 (as per IS 5831 1984)
- Voltage Grade: Upto and including 1100V AC 50Hz 3Ph
- Temperature: Max conductor temperature of 90° C
- Specification: Generally as per IS 7098 (Part 1) 1988

Sr. No.	PARAMETER		Size (Sq. mm) 2.5	4.0	6.0	10.0	16.0	25.0	35.0
1.	Conductor construction (No./mm)		36/0.3	56/0.3	84/0.3	140/0.3	226/0.3	354/0.3	495/0.3
2.	Conductor Resistance at 20° C (W*/km) Max.		7.41	4.95	3.3	1.91	1.21	0.780	0.554
3.	Insulation Thickness (mm)	Nom.	0.7	0.70	0.70	0.70	0.70	0.90	0.90
4.	Sheath Thickness (mm)	Nom.	1.0	1.0	1.1	1.2	1.3	1.5	1.6
5.	Overall Dimensions	Width(W)	12.7	14.9	16.9	20.3	23.8	29.6	33.6
	(mm) Approx.	Height (H)	6.0	6.6	7.4	9.1	10.4	12.9	14.4

The number of wires and its diameter in the conductor will be such as to satisfy requirement of the conductor resistance as per IS 8130: 1984.

Current Carrying Capacity (Amps)

Sr. No.	CABLE TYPE	Size (Sq. mm) 2.5	4.0	6.0	10.0	16.0	25.0	35.0
1.	PVC Insulation	18	26	31	42	57	72	90
2.	XLPE Insulation	30	37	46	66	85	113	139

Ambient Temperature (Deg.C)	25	30	35	40	45	50
Factor	1.18	1.12	1.06	1.00	0.94	0.88

Table 80: PERFORMANCE DATA OF SQUIRREL CAGE INDUCTION MOTORS (Source: M/S. Crompton Greaves Ltd.)

PERFORMANCE DATA

3 – PHASE TEFC SCR INDUCTION MOTOR

VOLTAGE : 415 ± 10% V	TYPE : SQUIRREL CAGE (SCR)	AMBIENT : 50 DEG C
FREQUENCY : 50 ± 5% Hz	RATING : CONTINUOUS TEMP	TEMR RISE (R) : 70 DEG C
COMBINED : ± 10% (ABSOLUTE	INSULATION : CLASS 'P' WITH CLASS 'b'	DEGREE OF : IPSS
VARIATION : SUM)	TEMR RISE	PROTECTION

OUTPUT		POLE	FRAME SIZE	FL RPM	FLC AMPS	FLT Kg-m	EFFICIENCY (%)			POWER FACTOR			DOL STG.T % FLT	STG STG C % FLC	POT % FLT	GO^2 KGM^2	NET WT kG
KW	HP						FL	3/4 LOAD	1/2 LOAD	FL	3/4 LOAD	1/2 LOAD					
0.18	0.25	2	AD63	2700	0.50	0.06	64.0	62.0	58.0	0.79	0.68	0.59	300	500	350	0.001	5.5
		4	AD63	1330	0.50	0.12	64.0	62.0	58.0	0.76	0.73	0.60	1200	400	250	0.003	5.6
		6	AD71	900	0.55	0.19	65.0	64.0	60.0	0.68	0.60	0.50	200	400	250	0.004	7.0
0.25	0.33	2	AD63	2700	0.65	0.09	58.0	67.0	63.0	0.79	0.72	0.60	275	500	325	0.001	5.6
		4	AD71	1350	0.75	0.18	62.0	61.0	55.0	0.76	0.73	0.60	225	400	275	0.004	7.0
0.37	0.50	2	AD71	2620	0.95	0.13	68.0	67.0	63.0	0.79	0.72	0.60	275	500	325	0.002	7.0
		4	AD71	1400	0.90	0.26	77.0	76.0	74.0	0.76	0.73	0.60	225	600	275	0.004	7.0
		6	AD/D80	910	1.05	0.40	69.0	68.0	66.0	0.71	0.63	0.52	210	400	260	0.011	10/17
		8	ADND90S	680	1.40	0.53	85.0	64.0	60.0	0.57	0.60	0.40	170	400	220	0.015	13.22
0.55	0.75	2	AD/D80	2820	1.30	0.19	73.0	72.0	68.0	0.81	0.73	0.62	250	600	300	0.003	10/17
		4	AD/D80	1410	1.25	0.38	77.0	76.0	74.0	0.78	0.75	0.64	200	500	250	0.007	10/17
		6	AD/D80	910	1.55	0.59	69.0	68.0	66.0	0.71	0.63	0.52	210	400	260	0.011	10/17
		8	ADND90S	680	1.75	0.79	69.0	68.0	64.0	0.63	0.54	0.45	150	400	200	0.021	13/22
0.75	1.00	2	AD/D80	2820	1.65	0.26	77.0	76.0	74.0	0.81	0.73	0.62	250	600	300	0.003	10/17
		4	AD/D80	1410	1.75	0.52	77.0	76.0	74.0	0.78	0.75	0.64	200	500	250	0.007	10/17
		6	AD/ND905	935	2.00	0.78	73.0	71.0	69.0	0.72	0.65	0.58	200	400	250	0.015	113/22
		8	AD/ND100L	700	2.55	1.04	70.0	69.0	64.0	0.58	0.51	0.41	200	400	250	0.003	19/32
1.10	1.50	2	AD/D80	2820	2.35	0.38	79.0	78.0	75.0	0.82	0.77	0.70	225	600	275	0.004	10/17
		4	AD/ND90S	1415	2.50	0.76	78.0	76.0	74.0	0.78	0.75	0.64	200	500	250	0.014	13/22
		6	AD/ND90L	935	2.85	1.15	74.0	73.0	70.0	0.72	0.65	0.58	200	400	250	0.021	16/25
		8	AD100L/ND100L	700	3.30	1.53	72.5	71.0	68.0	0.63	0.54	0.44	175	400	225	0.034	19/35

1.50	2.00	2	AD/ND90S	2830	3.20	0.52	80.0	79.0	77.0	0.82	0.77	0.70	225	600	275	0.006	13/22
		4	AD/ND90L	1415	3.20	1.03	80.0	79.0	77.0	0.81	0.78	0.71	200	500	250	0.019	16/25
		6	AD/ND100L	935	3.80	1.56	76.0	75.0	73.0	0.72	0.65	0.58	200	400	250	0.030	19/32
		8	AD112WND112M	700	4.00	2.09	77.0	77.0	75.0	0.68	0.60	0.52	190	400	240	0.057	29/45
2.20	300	2	AD/ND90L	28.30	4.60	0.76	81.0	80.0	78.0	0.82	0.77	0.70	300	600	350	0.008	16/25
		4	AD/ND100L	1440	4.60	1.49	82.0	81.0	79.0	0.82	0.78	0.72	200	600	250	0.030	19/32
		6	AD/ND112M	935	5.30	2.29	77.0	76.0	73.0	0.75	0.70	0.60	200	500	250	0.048	29/42
		8	ND132S	710	5.40	3.02	77.5	76.5	73.0	0.73	0.68	0.61	180	450	230	0.174	68.0
3.70	5.00	2	AD/ND100L	2875	7.60	1.25	83.0	82.0	80.0	0.82	0.77	0.70	250	650	300	0.022	19/36
		4	AD/ND112M	1440	7.60	2.50	84.0	83.0	81.0	0.81	0.76	0.69	200	600	250	0.052	29/42
		6	ND132S	950	7.70	3.79	85.0	84.0	82.0	0.79	0.73	0.63	200	600	250	0.174	68.0
		8	ND132M	710	8.60	5.08	80.0	80.0	78.0	0.75	0.70	0.60	180	600	230	0.214	79.0
5.50	7.50	2	AD/ND112M	2880	10.10	1.86	85.0	84.0	82.0	0.89	0.85	0.80	250	650	200	0.034	29/45
		4	ND123S	1450	10.10	3.69	85.0	85.0	83.0	0.88	0.85	0.75	250	600	300	0.131	68.0
		6	ND132M	950	1130	5.64	85.0	84.0	82.0	0.80	0.75	0.68	200	600	250	0.214	79.0
7.50	10.00	2	ND132S	2890	13.80	2.53	85.0	84.0	82.0	0.69	0.85	0.80	275	650	325	0.062	64.0
		4	ND132M	1455	13.60	5.02	87.0	86.0	84.0	0.88	0.85	0.75	275	600	324	0.161	79.0
9.30	12.50	2	ND132M	2890	16.70	3.13	87.0	86.0	84.0	0.89	0.85	280	275	600	325	0.076	79.0

NOTE: ALL PERFORMANCE FIGURES ARE SUBJECT TO TOLERANCES A6 PER IS325 # FOR 45°C/75°C ONLY.

PERFORMANCE DATA

3 – PHASE TEFC SCR INDUCTION MOTOR

VOLTAGE : 415 ± 10% V	TYPE : SQUIRREL CAGE (SCR)	AMBIENT : 50 DEG C
FREQUENCY : 50 ± 5% Hz	RATING : CONTINUOUS	TEMR RISE (R) : 70 DEG C
COMBINED : ± 10% OF (ABSOLUTE SUN VARIATION)	INSULATION : CLASS 'P' WITH CLASS 'b' TEMR RISE	DEGREE OF PROTECTION : IPSS

OUTPUT KW	OUTPUT HP	POLE	FRAME SIZE	FL RPM	FLC AMPS	FLT Kg-m	EFFICIENCY (%) FL	EFFICIENCY (%) 3/4 LOAD	EFFICIENCY (%) 1/2 LOAD	POWER FACTOR FL	POWER FACTOR 3/4 LOAD	POWER FACTOR 1/2 LOAD	DOL STG.T % FLT	STG STG C % FLC	POT % FLT	GO^2 KGM^2	NET WT kG
3.7	5.0	8	ND160M	710	8.0	5.08	83.0	83.0	81.0	0.74	0.70	0.62	150	500	200	0.46	125
5.5	7.5	6	ND160M	970	11.0	5.52	85.0	86.0	84.0	0.80	0.76	0.63	200	550	250	0.46	125
		8	ND160M	710	12.0	7.55	85.0	85.0	83.0	0.74	0.70	0.62	150	500	200	0.46	125
7.5	10.0	2	ND160M	2910	14.0	2.51	85.0	84.0	82.0	0.88	0.86	0.78	225	600	275	0.13	125
		4	ND160M	1460	14.0	5.00	87.0	87.0	85.0	0.85	0.81	0.73	225	600	275	0.31	125
		6	ND160M	975	15.0	7.49	87.5	87.0	85.0	0.80	0.76	0.68	200	550	250	0.46	125
		8	ND160L	710	16.0	10.29	85.0	85.0	83.0	0.76	0.72	0.64	150	500	200	0.64	148
9.3	12.5	2	ND160M	2920	17.0	3.10	87.0	37.0	85.0	0.88	0.88	0.78	250	650	300	0.13	125
		4	ND160M	1460	17.0	6.20	88.5	88.5	86.5	0.84	0.81	0.73	175	600	225	0.31	125
		6	ND160L	975	18.0	9.29	87.5	87.0	84.0	0.80	0.76	0.68	200	550	250	0.50	145
		8	ND189M	720 2	0.0	12.58	86.0	80.0	84.0	0.74	0.70	0.60	175	500	225	0.93	174
11	15	2	ND160M	2920	20.0	3.67	68.0	88.0	86.0	0.88	0.88	0.78	250	650	300	0.13	125
		4	ND160M	1480	21.0	7.34	89.0	89.0	85.0	0.82	0.79	0.70	200	600	250	0.36	125
		6	ND160L	975	29.0	14.98	90.0	90.0	88.0	0.79	0.73	0.66	225	600	275	1.16	210
		8	ND200L	725	24.0	14.88	82.0	87.0	85.0	0.74	0.70	0.60	179	500	225	1.16	210
15	20	2	ND160M	2920	26.0	5.00	89.5	87.5	0.88	0.80	0.80	0.79	250	650	300	0.17	125
		4	ND160L	1460	27.0	10.01	90.0	90.0	88.0	0.85	0.83	0.75	200	600	250	0.47	148
		6	ND180L	975	29.0	14.98	90.0	90.0	88.0	0.79	0.73	0.66	225	600	275	1.16	2.10
		8	ND200L	725	33.0	20.15	88.5	88.5	86.5	0.71	0.65	0.55	225	500	275	2.14	282
18.5	25	2	ND160L	2920	32.0	6.17	90.0	90.0	88.0	0.88	0.86	0.79	275	650	325	0.21	148
		4	ND180M	1475	33.0	12.22	92.0	92.0	90.0	0.94	0.90	0.72	200	600	250	0.81	174
		6	ND200L	975	34.0	18.48	91.0	91.1	89.9	0.84	0.82	0.73	200	550	250	1.69	282

		8	ND2255	725	39.0	24.85	89.0	89.0	97.0	0.75	0.71	0.63	175	500	225	3.24	345
22	30	2	ND180M	2930	41.0	7.31	91.0	91.0	89.0	0.83	0.80	0.72	225	600	275	0.44	164
		4	ND180L	1475	40.0	14.53	92.0	92.0	90.0	0.84	0.80	0.72	200	600	250	0.95	210
		6	ND200L	975	40.0	21.98	91.5	91.5	90.1	0.84	0.82	0.78	200	550	250	2.04	282
		8	ND225M	725	46.0	29.56	89.0	89.0	87.0	0.75	0.71	0.63	175	550	225	3.61	375
30	40	2	ND200L	2950	51.0	0.91	91.5	91.5	90.0	0.90	0.88	0.87	200	600	250	0.80	282
		4	ND200L	1475	51.0	19.81	92.0	92.0	90.2	0.89	0.86	0.78	290	600	300	1.62	282
		6	ND225M	980	53.0	29.82	92.0	92.0	90.5	0.85	0.82	0.73	200	550	250	3.61	379
		8	ND250M	735	61.0	39.76	90.5	90.5	88.5	0.75	0.71	0.83	175	550	225	4.82	473
37	50	2	ND200L	2950	62.0	12.22	92.0	92.0	90.0	0.90	0.89	0.87	200	600	250	0.89	282
		4	ND225M	1480	63.0	24.35	92.5	92.5	91.6	0.89	0.88	0.78	250	600	300	2.64	345
		6	ND250M	980	66.0	36.77	93.0	93.0	92.0	0.84	0.81	0.72	250	600	300	4.82	473
		8	ND280S	735	75.0	49.03	91.5	91.5	89.5	0.75	0.71	0.53	200	500	250	8.01	600
45	60	2	ND225M	2955	72.0	14.83	92.0	92.0	90.0	0.94	0.92	0.88	225	650	275	1.87	375
		4	ND225M	1490	76.0	29.61	93.0	93.0	91.5	0.89	0.86	0.78	250	600	300	3.13	275
		6	ND280S	990	79.0	44.72	93.0 9	3.0 9	1.0	0.85	0.92	0.73	250	550	300	8.01	600
		8	ND280M	725	91.0	60.46	92.0	92.0	90.5	0.75	0.71	0.63	175	500	225	9.89	670
55	75	2	ND250M	2955	58.0	18.13	92.5	92.5	90.5	0.94	0.92	0.88	175	600	225	2.79	473
		4	ND250M	1475	92.0	36.32	93.5	93.5	92.0	0.89	0.85	0.82	200	600	250	3.45	473
		6	ND280M	980	95.0	54.66	93.5	93.5	92.5	0.86	0.82	0.74	176	600	225	9.89	670
		8	ND3155	740	113.0	72.39	93.0	93.0	91.5	0.73	0.65	0.56	250	550	300	14.12	900
75	100	2	ND280S	2975	124.0	24.55	93.5	92.0	0.90	0.86	0.78	225	600	275	7.14	600	
		4	ND280S	1480	123.0	49.36	94.0	94.0	92.5	0.90	0.88	0.84	250	600	300	7.21	600
		6	ND315S	987	129.0	74.01	94.0	94.0	92.5	0.86	0.82	0.74	250	600	300	14.12	900
		8	ND315M	740	153.0	98.72	93.5	93.5	91.5	0.73	0.66	0.56	200	250	18.98	950	

NOTE: ALL PERFORMANCE FIGURES ARE SUBJECT TO TOLERANCES A6 PER IS325 # FOR 45G°C/75°C ONLY.

Table 81: PERFORMANCE DATA OF SLIP-RING INDUCTION MOTORS
(Source: M/s. Crompton Greaves Ltd.)

PERFORMANCE DATA

3 – PHASE TEFC SCR INDUCTION MOTOR

VOLTAGE	: 415 ± 10% V	TYPE	: SQUIRREL CAGE (SCR)	AMBIENT	: 45 DEG C
FREQUENCY	: 50 ± 5% Hz	RATING	: CONTINUOUS	TEMR RISE (R)	: 75 DEG C
COMBINED VARIATION	: ± 10% (ABSOLUTE SUM)	INSULATION	: CLASS 'P'	REFERENCE	: IS 325, BS4999 & RELEVANT STDS
		PROTECTION	: IPSS	DUTY S1	

OUTPUT KW	OUTPUT HP	POLE	FRAME SIZE	FL RPM	FLC AMPS	EFFICIENCY (%) FL	EFFICIENCY (%) 3/4 LOAD	EFFICIENCY (%) 1/2 LOAD	POWER FACTOR FL	POWER FACTOR 3/4 LOAD	POWER FACTOR 1/2 LOAD	OC ROTOR VOLTS	ROTOR CURRENT AMPS	GD^2 KGM^2	NET WT
3.0	4.0	8	NDW160M	690	8.0	80.0	79.0	77.0	0.68	0.64	0.56	230	8.0	0.58	141
3.7	5.0	8	NDW160M	710	9.0	81.0	80.0	78.0	0.68	0.64	0.56	230	10.0	0.58	141
5.5	7.5	6	NDW160M	940	12.0	83.0	82.0	80.0	0.76	0.74	0.66	180	18.5	0.58	141
		8	NDW160L	710	18.0	84.0	83.0	81.0	0.65	0.64	0.56	350	13.0	1.58	235
7.5	10.0	4	NDW160M	1440	15.0	83.0	82.0	80.0	0.84	0.80	0.72	350	13.0	0.48	141
		6	NDW160L	940	17.0	83.0	82.0	80.0	0.76	0.72	0.64	260	19.0	0.77	165
		8	NDW180L	710	18.0	84.0	83.0	81.0	0.68	0.64	0.56	350	13.0	1.58	235
93	12.5	4	NDW160L	1440	18.0	84.0	83.0	81.0	0.84	0.8	0.72	350	16.0	0.35	165
		6	NDW180M	950	18.0	86.0	85.0	83.0	0.82	0.78	0.70	230	24.0	1.33	205
		8	NDW200L	725	22.0	86.0	85.0	83.0	0.69	0.64	0.76	300	19.5	2.68	313
11.0	15.0	4	NDW160L	1440	22.0	84.0	83.0	81.0	0.84	0.8	0.72	350	16.0	0.35	165
		6	NDW180L	950	22.0	86.0	85.0	83.0	0.82	0.78	0.70	230	28.5	1.58	235
		8	NDW200L	725	26.0	87.0	86.0	84.0	0.68	0.64	0.56	300	23.0	2.68	313
15.0	20.0	4	ND2180L	1445	28.0	88.0	87.0	85.0	0.84	0.8	0.72	350	27.0	1.26	235
		6	NDW225L	970	29.0	89.0	88.0	86.0	0.82	0.78	0.70	295	31.0	2.68	313
		8	NDW225M	730	34.0	87.0	86.0	84.0	0.7	0.66	0.58	350	27.0	3.06	410
18.5	25.0	4	NDW200L	1470	34.0	89.0	88.0	86.0	0.84	0.80	0.72	300	36.0	2.18	313
		6	NDW225M	970	35.0	89.0	88.0	86.0	0.82	0.78	0.70	310	37.0	2.87	410
		8	NDW225M	730	42.0	88.0	87.0	85.0	0.70	0.66	0.58	360	32.0	3.63	410
22.0	30.0	4	NDW200L	1470	40.0	89.0	88.0	86.0	0.85	0.81	0.73	390	33.0	2.34	313

		6	NDW225M	970	42.0	89.0	88.0	85.0	0.82	0.78	0.70	310	37.0	2.87	410
		8	NDW225M	730	45.0	91.0	90.0	83.0	0.74	0.70	0.62	250	55.0	6.81	513
30.0	40.0	4	NDW225M	1470	55.0	89.0	88.0	85.0	085	0.81	0.73	380	47.0	2.96	410
		6	NDW250M	975	56.0	91.0	90.0	88.0	0.82	0.78	0.70	320	56.0	6.81	513
		8	NDW280S	735	62.0	91.0	90.0	88.0	0.74	0.70	0.62	440	43.0	12.89	650
37.0	50.0	4	NDW250M	1475	67.0	91.0	91.0	89.0	0.85	0.81	0.73	425	52.0	4.96	513
		6	NDW280S	980	67.0	91.0	90.0	88.0	0.84	0.80	0.72	410	57.0	12.89	650
		8	NDW280M	735	76.0	91.0	90.0	88.0	0.74	0.70	0.62	520	45.0	15.14	720
45.0	60.0	4	NDW250M	1475	82.0	90.0	89.0	87.0	0.85	0.81	0.73	445	60.0	5.70	513
		6	NDW280M	980	80.0	91.5	91.0	89.0	0.85	0.81	0.73	320	87.0	15.14	720
		8	NDW315S	735	112.0	91.0	90.0	88.0	0.74	0.70	0.62	320	88.0	21.65	950
55.0	75.0	4	NDW280S	1480	97.0	91.5	91.0	89.0	0.86	0.82	0.74	460	73.0	10.61	650
		6	NDW315S	980	98.0	92.0	91.0	89.0	0.85	0.81	0.73	320	87.0	15.14	720
		8	NDW315M	735	112.0	91.0	90.0	88.0	0.75	0.71	0.63	320	108.0	24.16	1000
75.0	100	4	NDW280M	1480	130.0	92.5	92.0	90.0	0.87	0.83	0.75	490	92.0	12.45	720
		6	NDW315M	980	132.0	93.0	92.0	90.0	0.85	0.81	0.73	470	96.0	24.16	1000
		8	NDW315L	730	146.0	93.0	92.0	90.0	0.77	0.72	0.64	357	145.0	33.32	1000
90.0	120.0	4	NDW315S	1485	1570	92.5	92.0	90.0	0.86	0.82	0.74	505	106.0	18.22	950
		6	NDW315M	980	169.0	92.5	92.5	90.0	0.80	0.76	0.66	357	140.0	24.26	1000
110.0	150.0	4	NDW315M	1485	192.0	92.5	92.0	90.0	0.86	0.82	0.74	500	132.0	20.21	1000
		6	NDW315L	980	191.0	93.0	92.0	90.0	0.86	0.82	0.74	400	165.0	34.52	1220
		8	NDW355L	730	234.0	91.0	90.0	88.0	0.72	0.70	0.62	450	145.0	51.05	1260
132.0	180.0	6	NDW315L	980	222.0	94.0	93.0	91.0	0.88	0.82	0.69	340	220.0	34.52	1220
		8	NDW355LX	740	277.0	92.0	92.0	88.0	0.72	0.70	0.62	505	150.0	51.05	1260
160.0	215.0	4	NDW315L	1480	271.0	93.5	92.0	89.0	0.88	0.82	0.69	440	220.0	34.52	1220
		6	NDW355LX	985	275.0	94.0	94.0	91.0	0.86	0.82	0.74	415	235.0	43.49	1260
225.0	300.0	4	NDW355LX	1485	387.0	94.0	94.0	92.0	0.86	0.83	0.74	525	290.0	29.56	1260
250.0	335.0	4	NDW355LX	1485	428.0	94.5	94.5	92.0	0.86	0.83	0.74	525	330.0	29.56	1260

FL = Full load: FLC Load Current; OC = Open Circuit Pull Out Torque 275% FLT.

NOTE: ALL PERFORMANCE FIGURES ARE SUBJECT TO TCLERANCES AS PE4R IS325

Table 82: PERFORMANCE DATA OF ENERGY EFFICIENT MOTORS
(Source: M/s. Kirloskar Electricals Ltd.)

	PERFORMANCE FIGURES – 415 VOLTS, 50HZ 50° C AMB. 10-5-10 VARIATIONS												
	Frame	Output (kw)	F.L.S. (rpm)	F.L.C. (A)	L.R.C. (p.u.)	L.R.T. (p.u.)	P.O.T. (p.u.)	% Efficiency			Power Factor		
								FL	3/4FL	1/2FL	FL	3/4FL	1/2FL
2 POLE	180 M	22	2945	37	6.5	1.8	2.5	91.0	90.0	88.0	0.90	0.88	0.86
	200 L	30	2955	50	7.5	2.2	2.5	91.0	90.0	88.0	0.91	0.90	0.66
	200 L	37	2955	62	7.5	2.2	2.5	91.0	90.0	88.0	0.91	0.90	0.86
	225 M	45	2965	75	7.0	2.2	2.5	91.0	90.0	88.0	0.92	0.90	0.87
	250 M	55	2965	92	6.0	1.6	2.8	92.5	92.0	90.0	0.90	0.88	0.80
	280 S	75	2970	125	6.3	1.7	2.9	93.0	92.5	90.5	0.90	0.87	0.80
	280 M	90	2970	149	6.0	1.6	2.8	93.5	93.0	91.0	0.90	0.87	0.80
	315 S	110	2978	181	6.6	1.7	3.2	94.0	93.0	92.0	0.90	0.86	0.78
	315 M	125	2975	205	6.0	1.5	2.8	94.5	94.0	92.0	0.90	0.87	0.80
	315 M	135	2978	221	6.5	1.6	3.0	94.5	94.0	92.0	0.90	0.87	0.80
	315 L	150	2978	243	7.0	1.6	3.0	94.5	94.0	92.0	0.91	0.89	0.83
	315 L	160	2980	258	7.0	1.7	3.0	95.0	94.5	93.0	0.91	0.89	0.83
	315 L	180"	2978	289	6.3	1.5	2.7	95.2	94.8	93.5	0.91	0.89	0.84
4 POLE	180 M	18.5	1470	33	6.5	2.5	2.75	91.0	90.5	90.0	0.87	0.83	0.75
	180 L	22	1470	38	6.5	2.5	2.75	91.5	91.0	90.0	0.87	0.83	0.75
	200 L	30	1475	50	6.5	2.2	2.75	92.0	91.0	90.0	0.90	0.89	0.81
	225 S	37	1475	62	7.5	2.5	2.75	92.0	91.5	90.0	0.90	0.86	0.77
	255 M	45	1475	75	7.5	2.5	2.75	92.5	91.5	90.0	0.90	0.86	0.77
	250 M	55	1480	94	6.5	2.4	2.7	93.5	93.0	91.5	0.87	0.84	0.75
	280 S	75	1483	128	6.5	2.2	2.6	94.0	93.5	92.0	0.87	0.84	0.75
	280 M	90	1845	153	6.8	2.5	2.8	94.5	94.0	92.5	0.87	0.84	0.75
	315 S	110	1483	186	6.5	2.5	2.7	94.5	94.0	29.5	0.87	0.84	0.76
	315 M	125	1482	212	6.5	2.2	2.7	94.5	94.0	92.5	0.87	0.85	0.80
	315 M	135	1483	225	6.5	2.5	2.7	95.0	94.5	93.0	0.88	0.86	0.80
	315 L	150	1484	250	7.0	2.5	2.7	95.0	94.5	93.0	0.88	0.86	0.80
	315 L	160	1483	266	6.6	2.5	2.7	95.2	94.8	93.5	0.88	0.86	0.80
	315 L	180"	1483	299	6.5	2.4	2.5	95.2	95.0	94.0	0.88	0.87	0.81

6	180 L	15	970	29	6.0	1.8	2.3	89.0	89.0	88.5	0.82	0.78	0.74
	200 L	10.5	970	34	6.0	2.0	2.5	90.5	90.5	88.5	0.84	0.80	0.72
	200 L	22	970	40	6.0	2.0	2.5	90.5	90.5	88.5	0.84	0.80	0.72
	225 M	30	970	53	6.0	2.0	2.3	91.0	90.5	88.5	0.86	0.83	0.78
P	250 M	37	983	66	6.0	1.8	2.5	92.5	91.5	89.0	0.85	0.81	0.74
	280 S	45	985	80	6.3	2.2	2.5	93.0	92.0	90.0	0.84	0.80	0.70
O	280 M	55	985	97	6.3	2.2	2.5	93.5	92.5	90.5	0.85	0.81	0.72
L	315 S	75	988	131	6.5	2.5	2.8	93.8	93.0	91.0	0.85	0.81	0.72
	315 M	90	988	157	6.6	2.5	2.8	94.0	93.5	92.0	0.85	0.81	0.72
E	315 M	110	968	191	6.6	2.5	2.8	94.3	93.8	92.2	0.85	0.81	0.71
	315 L	125	987	217	6.5	2.4	2.7	94.5	94.0	92.5	0.85	0.82	0.73
	315L	135"	988	234	6.7	2.6	2.8	94.5	94.0	92.5	0.85	0.81	0.72
	180 L	11	725	24	6.0	1.8	2.3	88.0	89.0	87.0	0.74	0.68	0.57
8	200 L	15	730	30	6.0	2.0	2.3	89.0	89.0	88.0	0.78	0.74	0.64
	225 S	18.5	735	37	6.0	2.0	2.3	90.0	90.0	88.0	0.78	0.74	0.64
	225 M	22	735	44	6.0	2.0	2.3	90.0	90.0	88.0	0.78	0.74	0.64
	250 M	30	736	59	5.5	1.7	2.5	91.5	90.5	87.0	0.78	0.70	0.58
P	280 S	37	738	73	6.0	2.0	2.4	92.0	91.0	88.0	0.77	0.70	0.57
	280 M	45	740	89	6.0	2.0	2.S5	92.0	91.0	88.0	0.77	0.70	0.57
O	315 S	55	740	152	6.0	2.0	2.5	92.5	91.5	88.5	0.78	0.72	0.60
	315 M	75	740	152	6.0	2.0	2.5	92.8	91.7	89.0	0.74	0.67	0.56
L	315 L	90	740	182	6.0	2.1	2.5	93.0	92.0	89.0	0.75	0.68	0.57
E	315 L	110"	738	220	6.0	2.0	2.3	93.0	92.5	90.5	0.75	0.68	0.57

Table 83: PERFORMANCE DATA OF ENERGY EFFICIENT MOTORS (8 POLE)
(Source: M/s. Havells India Ltd.)

Insulation class F with temperature rise limited to class B

Type		Rated output kW	HP	Rated speed RPM	Efficiency 50%	75%	100%	Power factor cos φ	Rated current at 415 V I_x	Direct-on-line starting: starting current ratio I_A/I_x	Breakaway torque ratio M_A/M_x	Pull-up torque ratio M_A/M_x	Pull-out torque ratio M_A/M_x	Moment of inertia J 10^3 kgm^2	Weight kg
750 RPM (8 Pole)														Aluminium	Frame
MH 71Z AA	8	0.12	0.16	570	47.7	51.7	52	0.55	0.58	2.4	2.5	2.4	2.5	0.90	6.3
MH BOZ	8	0.25	0.33	680	51.4	54.4	55	0.65	0.97	2.2	1.8	1.9	2.0	1.97	8.0
MH 90S AA	8	0.37	0.50	680	58.2	64.2	62.1	0.57	1.45	2.1	1.4	1.3	1.6	3.18	11.4
MH 90L BA	8	0.55	0.75	680	58.2	64.2	62.1	0.58	2.12	2.1	1.4	1.3	S1.6	4.78	15.0
MH 100L AA	8	0.75	1.0	690	64.3	69.3	70.3	0.68	2.18	3.0	1.6	1.5	1.7	6.72	17.6
MH 100L BA	8	1.1	1.5	690	63.8	71.8	72	0.68	3.13	3.0	1.9	1.3	1.4	15.93	22.6
MH 112M AA	8	1.5	2.0	696	70.5	73.5	74.5	0.69	4.06	4.0	1.8	2.0	2.4	16.7	35
MH 132S ZA	8	2.2	3.0	710	82.5	83.7	82.2	0.69	5.40	3.4	1.7	1.6	1.7	29.5	45.5
MH 132M ZA	8	3	4.0	710	84.3	85.2	83.4	0.73	6.86	3.6	1.7	1.6	2.1	37.7	54.4
MH 160M YA	8	3.7	5.0	700	86.2	86.1	85.7	0.77	7.80	4.5	1.8	1.6	2.3	09.5	795
MH 160M ZA	8	5.5	7.5	720	86.7	86.7	86.3	0.77	11.52	4.0	1.8	1.6	2.3	119.5	90.5
MH 160L ZA	8	7.5	10	710	87.4	86.1	86.9	0.78	15.39	4.0	1.8	1.6	2.3	150.2	96
750 RPM (8 Pole)														Cast Iron	Frame
MH 180L ZG	8	11	15	725	87.7	88.8	87.9	0.78	22.3	4.6	2.1	1.4	1.9	215	175
MH 200L RG	8	S15	20	730	88.1	89.7	89.4	0.78	29.9	5.3	2.3	1.9	2.5	285	220
MH 225S P	8	18.5	25	730	89.4	90.6	90.3	0.79	36.1	5.2	2.3	1.9	2.2	438	255
MH 225M P	8	22	30	730	89.5	90.7	90.3	0.79	42.9	5.6	2.5	2.0	2.3	538	285
MH 250M P	8	30	40	730	89.4	90.9	90.9	0.80	57.4	6.0	2.8	2.5	2.6	1000	400
MH 280S V	8	37	50	740	91.7	92.9	92.8	0.83	66.8	6.0	2.1	1.7	2.3	1520	480
MH 280M V	8	45	60	740	91.7	93	92.9	0.83	81.2	6.0	2.1	1.7	2.3	1870	500
MH 315S YE	8	55	75	738	92.6	93.4	93.1	0.84	97.8	6.0	2.1	1.7	2.3	2200	620
MH 315M YE	8	75	100	738	93.5	94.3	94.1	0.84	132.0	6.0	2.1	1.8	2.1	3100	790
MH 315M ZE	8	90	125	738	93.7	94.5	94.3	0.84	158.1	6.0	2.2	1.8	2.1	3600	900
MH 315L ZE	8	110″	150	738	93.8	98.8	94.4	0.84	193.0	6.0	2.5	1.8	2.1	4300	990

″ Temperature rise to class F

Note: Performance figures are subject to IS tolerance

Table 84: PERFORMANCE DATA OF ENERGY EFFICIENT MOTORS (4 POLE)
(Source: M/s. Havells India Ltd.)

Three-phase motors designed
for range of rated voltage
415 V ± 10% - 50 Hz ± 5%

EFFZ

Insulation class F with
temperature rise limited to class B

Type	Rated output		Rated speed	Efficiency			Power factor	Rated current at 415 V	Direct-on-line starting: starting current ratio	Breakaway torque ratio	Pull-up torque ratio	Pull-out torque ratio	Moment of inertia	Weight
							cos ϕ	I_x	I_A/I_x	M_A/M_x	M_A/M_x	M_A/M_x	J	
	kW	HP	RPM	50%	75%	100%							10^3 kgm^2	kg
													Aluminium	Frame
1000 RPM (6 Pole)														
MH 712 AA 6	0.18	0.25	850	53	56	52	0.72	0.67	2.2	1.6	1.5	1.8	0.6	5.7
MH 712 BA 6	0.25"	0.33"	870	54	58	60	0.7	0.83	2.5	1.7	1.6	1.7	0.9	6.3
MH BOZ AA 6	0.37	0.50	910	55	66	64	0.73	1.10	2.7	1.6	1.6	2.1	1.97	8
MH BOZ BA 6	0.55	0.75	910	63	67	70	0.72	1.52	2.9	2.2	2.1	2.1	2.47	9.4
MH 90S AA 6	0.75	1.0	910	73	75	73	0.72	1.99	2.9	1.7	1.5	1.7	3.18	11.6
MH 90L BA 6	1.1	1.5	908	74	76	75	0.72	2.83	3.0	1.7	1.5	1.7	4.78	15
MH 100L AA 6	1.5	2.0	930	74.5	79.5	76	0.73	3.76	3.7	1.8	1.6	2.3	6.73	17.5
MH 100L BA 6*	1.8	2.5	940	73	77	78	0.70	4.59	4.2	2.4	2.4	2.8	9.43	22
MH 112M AA 6	2.2	3.0	940	82.4	84.4	83.4	0.7	4.83	4.4	2.4	2.4	2.6	14.18	20
MH 112M BA 6	3	4.0	940	84.2	85.2	85.2	0.77	6.36	5.3	2.9	2.9	2.9	18.7	39
MH 132S ZA 6	3	4.0	950	82.8	84.9	84.6	0.73	6.73	4.9	2.1	1.8	2.4	23.5	3.7
MH 132M YA 6	3.7	5.0	950	84.4	86.1	85.7	0.76	7.90	4.5	2.3	2.0	2.5	29.5	42.5
MH 132M ZA 6	5.5	7.5	950	85	86.4	85.8	0.76	11.73	4.1	2.3	1.9	2.2	37.3	55.5
MH 132M TA 6	7.5	10	950	66.1	86.8	86.1	0.7	15.95	5.0	2.3	1.9	2.8	54.1	64.1
MH 160M ZA 6	7.5	10	970	87.8	88.7	88	0.76	15.60	3.9	1.9	1.8	1.6	81.2	99
MH 160L ZA 6	11	15	960	89.5	89.9	89	0.34	20.47	5.0	2.5	2.2	3.5	105.7	113.6

Table Contd.

1000 RPM (6 Pole) Cast iron frame															
MH 180L ZG	6	15	20	970	91.3	91.6	90.8	0.84	27.4	6.7	2.2	1.8	2.8	169	155
MH 200L PG	6	18.5	25	970	90.6	91.9	91.7	0.84	33.4	5.3	2.2	1.8	2.3	260	210
MH 200L RG	6	22	30	975	91.3	92.4	92.2	0.84	39.5	5.7	2.2	1.8	2.3	285	220
MH 225M P	6	30	40	975	92	93	92.8	0.84	53.5	5.7	2.3	1.6	2.3	536	290
MH 250M P	6	37	50	975	92	93.1	93	0.85	65.1	6.0	2.8	2.5	2.6	880	380
MH 280S V	6	45	60	985	92.5	93.8	93.5	0.87	77.0	5.6	1.8	1.5	2.4	1350	460
MH 280M V	6	55	75	985	92.5	93.7	93.6	0.87	94.0	5.6	1.8	1.5	2.4	140	515
MH 315S YE	6	75	100	985	93.5	94.3	94.3	0.87	127.2	6.0	1.8	1.7	2.6	2200	20
MH 315M YE	6	90	125	988	94.5	95.1	94.8	0.88	150.1	6.0	2.0	2.0	2.6	3100	790
MH 315M ZE	6	110	150	987	94.8	95.3	95	0.88	183.1	6.0	2.0	2.0	2.6	3600	860
MH 315L ZE	6	13228	18008	987	95.3	95.6	95.2	0.89	216.7	6.0	2.0	2.0	2.5	4300	990

*Temperature rise to class F

*Higher output (progressive motor)

Note: Performance figures are subject to IS tolerance

Table 85: PERFORMANCE DATA OF ENERGY EFFICIENT MOTORS (4 POLE)
(Source: M/s Havells India Ltd.)

Insulation class F with temperature rise limited to class B

Type		Rated output		Rated speed	Efficiency			Power factor	Rated current at 415 V	Direct-on-line starting current ratio	Direct-on-line starting Breakaway torque ratio	Direct-on-line starting Pull-up torque ratio	Pull-out torque ratio	Moment of inertia	Weight
		kW	HP	RPM	50%	75%	100%	cos φ	I_x	I_A/I_x	M_A/M_x	M_A/M_x	M_A/M_x	J 10^3 kgm^2	kg
1500 RPM (4 Pole)														Aluminium Frame	
MH 56Z AA	4	0.06	0.08	1300	45.9	47.9	51.9	0.58	0.28	2.8	2.1	2.0	2.1	0.14	3.2
MH 56Z BA	4	0.09	0.12	1330	46.7	50.7	54.7	0.66	0.35	2.5	2.2	2.1	2.2	0.14	3.3
MH 63Z AA	4	0.12	0.16	1350	49.5	53.5	57.5	0.63	0.46	2.4	2.0	1.9	2.0	0.25	4.1
MH 56Z BA	4	0.18	0.25	1330	55.2	58.2	64.2	0.7	0.56	2.3	1.9	1.8	1.9	0.27	4.6
MH 63Z CA	4	0.25	0.33	1360	55.3	58.8	64.3	0.76	0.71	2.7	2.2	2.0	2.1	0.30	4.9
MH 71Z AA	4	0.25	0.33	340	57.7	61.7	66.7	0.65	0.80	3.2	1.9	1.8	2.0	0.63	5.2
MH 71Z BA	4	0.37	0.5	1370	65	68	72	0.71	1.01	3.3	2.2	2.1	2.2	0.76	5.4
MH 71Z CA	4	0.55"	0.75"	1380	65.8	68.3	72.8	0.75	1.40	3.6	2.4	2.3	2.4	0.98	6.3
MH 80Z AA	4	0.5	50.75	1400	71.5	73.5	74.5	0.78	1.32	3.6	2.6	2.5	2.6	1.58	8.2
MH 80Z BA	4	0.75	1.0	1410	66.9	73.4	75	0.73	1.91	4.4	2.8	2.3	2.6	2.00	9.3
MH 80Z CA	4	1.1"	1.5"	1385	75.9	78.2	77.7	0.81	2.43	4.4	2.5	2.9	2.6	2.41	10.6
MH 90S AA	4	1.1	1.5	1400	71.4	77.2	78.3	0.82	2.38	5.2	2.5	2.4	2.8	2.5	12.5
MH 90L BA	4	1.5	2.0	1400	77.2	80.3	60.2	0.81	3.21	5.7	2.8	2.6	3.0	3.13	14.5
MH 90L CA	4	1.8"	2.5"	1380	78.5	81.2	80.7	0.83	3.74	5.5	2.7	2.5	2.9	3.13	14.5
MH 90L DA	4	2.2"	3.0"	1400	79.4	82.4	62.4	0.82	4.53	4.8	2.9	2.8	3.2	4.06	17
MH 100L AA	4	2.2	3.0	14.35	79.5	82.2	83	0.78	4.73	5.3	2.5	2.4	2.7	4.6	19.5
MH 100L BA	4	3	4.0	1425	83	84.7	84.1	0.8	6.20	4.6	2.4	2.3	2.5	5.58	22.5
MH 100L CA	4	3.7	5.0"	1400	83.5	84.4	83	0.82	7.56	6.0	2.6	2.4	2.9	6.05	25
MH 112M AA	4	3.7	5.0	1430	85.7	86.4	85.4	0.84	7.18	8.0	2.0	2.0	2.8	12.2	29.6
MH 112M BA	4	5.5"	7.5"	1430	87	87.3	86.3	0.86	10.31	6.0	2.0	2.0	2.9	15.2	34
MH 132S ZA	4	5.5	7.5	1430	86.7	87.7	86.8	0.86	10.25	5.8	33.0	2.7	2.9	22	41.9
MH 132M ZA	4	7.9	10	1440	88.7	89.1	88	0.88	13.47	6.0	2.6	2.7	2.9	30	51
MH 132M ZA	4	9.3	12.5"	1440	88.3	89.1	88.2	0.88	16.67	6.0	2.75	3.0	3.5	37.2	65

MH 132M TA 4	11"	15"	1440	88.4	88.8	89.4	0.88	19.45	6.0	3.0	3.3	3.5	37.2	65
MH 160M XA 4	11	15	1460	89.3	90.1	89.5	8.83	20.6.	6.0	2.1	2.2	2.9	81.2	88.5
MH 160L XA 4	15	20	1460	90.4	91.1	90.4	0.88	26.23	6.0	2.1	2.2	3.5	105.7	106.5
1500 RPM (4 Pole)													Cast irom frame	
MH 180M XG 4	18.5	25	1460	91.7	92.5	92	0.87	32.2	6.0	2.2	2.2	3.0	105	150
MH 180L XG 4	22	30	1460	91.9	92.6	92	0.87	38.2	6.0	2.2	2.2	3.0	118	160
MH 180L RG 4*	30"	40	1455	91.7	92.7	92.7	0.85	52.9	6.0	2.5	2.4	3.2	150	175
MH 2001L NG 4	30	40	1468	91.8	92.9	92.8	0.87	51.7	6.0	2.0	1.8	2.6	195	225
MH 200L FG 4*	37"	50"	1465	92.5	93.2	93.2	0.86	64.2	6.0	2.2	2.0	2.5	248	255
MH 225S N 4	37	50	1475	91.4	92.7	92.8	0.87	63.8	5.9	2.75	2.1	2.9	256	290
MH 225M N 4	45	60	1475	92.3	93.3	93.2	0.88	76.3	6.0	2.0	2.0	2.9	461	330
MH 250M N 4	55	75	1475	93.3	94	93.8	0.87	93.8	6.0	2.0	2.1	2.4	640	385
MH 250M K 4*	75	100	1470	92.6	93.7	93.9	0.85	130.7	6.0	3.0	2.3	2.7	812	440
MH 280S T 4	75	100	1480	92.7	93.9	94	0.87	127.6	6.0	1.8	1.8	2.7	990	510
MH 280M T 4	90	125	1480	93.5	94.6	94.7	0.87	152.0	6.0	1.8	1.8	2.7	1200	565
MH 315S YE 4	110	150	1480	94.1	95	96	0.87	185.2	6.0	1.8	1.8	2.8	1400	620
MH 315M YE 4	132	180	1485	94.3	95.4	95.7	0.88	218.1	6.0	1.8	1.6	2.8	2200	860
MH 315M ZE 4	160	220	1485	94.6	95.7	95.9	0.9	257.9	6.0	1.5	1.5	2.8	2500	940
MH 3751M YE4	200"	270"	1485	95.2	96	96.2	0.91	317.8	6.0	1.8	1.6	2.8	3100	1120

"temperature rise limited to class B

*Higher output progressive motor)

Note : Performance figures are subject to IS tolerance

Table 86: DUTY TYPES OF MOTORS (REFERENCE: IS-325)

Duty type symbol	Duty type	Applications
S1	Continuous duty	Pumps, blowers, compressors, fans etc.
S2	Short time duty	Siren, capstan, flood relief gates, etc.
S3	Intermittent periodic duty	Valve actuators, wire drawing machine
S4	Intermittent periodic with starting	Hoists, cranes, lifts, escalators
S5	Intermittent periodic with starting / braking	Hoists, cranes (with electric braking) rolling mills
S6	Continuous duty with intermittent periodic loading	Machine tools, conveyors
S7	Continuous duty with starting and braking	Machine tools
S8	Continuous duty with periodic load changes	Pole changing applications

Table 87: DE-RATING OF MOTOR CAPACITY FOR AMBIENT TEMPERATURE

Cooling air temperature °C	Permissible output %
30	107
35	104
40	100
45	95
50	90
55	83

Table 88: DE-RATING OF MOTOR CAPACITY FOR ALTITUDE

Altitude-meters	Permissible output %
1000	100
2000	95
3000	90
4000	85

Table 89: PERMISSIBLE TEMPERATURES OF MOTORS (Reference-IS-325)

Class of insulation	Permissible temperature rise from 40ºC	Hot spot temperature T_{max}	Average insulation life at /below rated temperature-hours
Class B	70ºC	130ºC	20000
Class F	95ºC	155ºC	20000
Class H	115ºC	180ºC	20000

Table 90: CAPACITOR RATING FOR DIRECT CONNECTION TO INDUCTION MOTORS

Motor rating HP/kW	Capacitor rating for motor speed in rpm (kVAR)						Cable size mm^2
	3000	1500	1000	750	600	500	
3/2.5	1	1	1.5	2	2.5	2.5	4
5/3.7	2	2	2.5	3.5	4	4	4
7.5/5.7	2.5	3	3.5	4.5	5	5.5	6
10/7.5	3	4	4.5	5.5	6	6.5	6
15/11.2	4	5	6.7	7.5	8.5	9	10
20/15	5	6	7	9	11	12	16
25/18.7	6	7	9	10.5	13	14.5	25
30/22.8	7	8	10	12	15	17	25
50/37	11	12.5	16	18	23	25	50
75/57	16	17	21	23	29	32	95
100/75	21	17	21	23	29	32	185
130/102	31	33	36	38	48	55	185
200/150	40	42	45	47	60	67	185
250/187	46	50	53	55	68	76	240

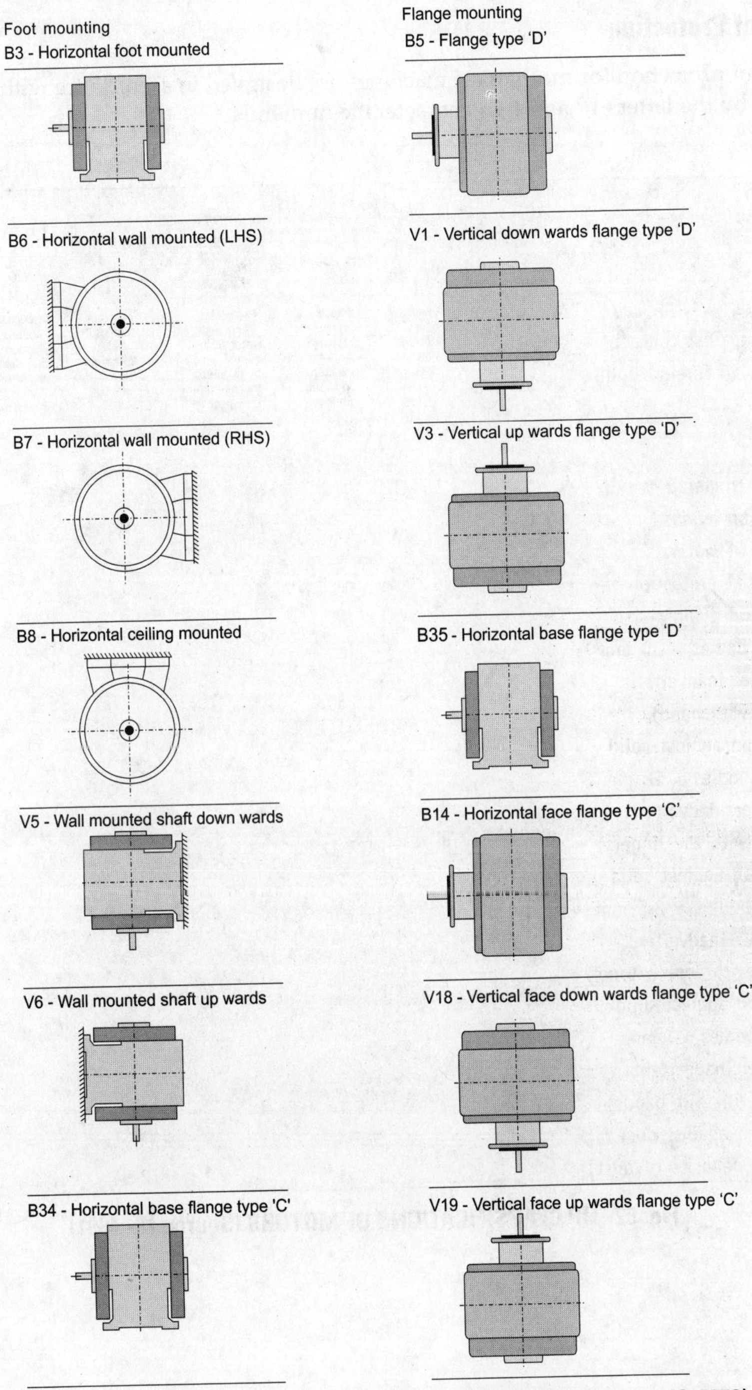

Fig. 91 STANDARD TYPES OF MOUNTING OF MOTORS (Source: IS-2253)

Degree of Protection

Degrees of protection for mechanical machines are designed in accordance with IS 4691/IEC 600034.5 by the letters IP and two characterstic numerals

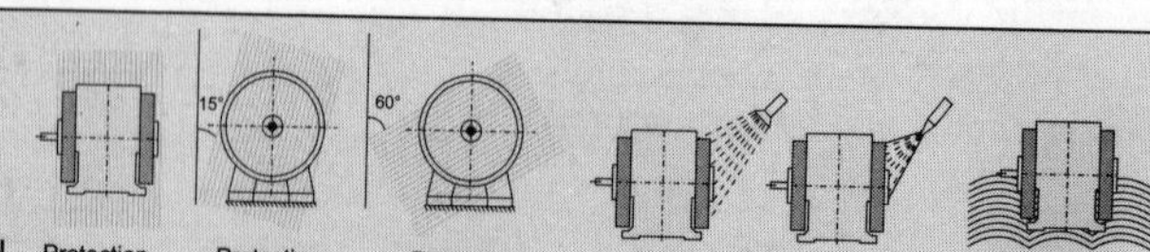

First numeral Protection against contact and ingress of foreign bodies	No. special protection	Protection against vertically falling water drop	Protection against droppings water when inclined by up to 15 degrees	Protection against water spray when inclined by up to 60 degree from vertical	Protection against water splashed from any direction	Protection against water projected by noozle from any direction	Protection against heavy seats
Second Numeral: Protection against → ingress of water	0	1	2	3	4	5	6
0 No special protection							
1 Protection against sold foreign bodies > 50 mm (Example: inadvertent contact with hand)							
2 Protection against solid foreign bodies > 12 mm (Example: inadvertent contact with the fingers)		IP 21	IP 22	IP 23			
3 Protection against solid foreign bodies > 2.5 mm (Example: Inadvertent contact with wire & tools)							
4 Protection against solid foreign bodies > 1 mm (Example: Inadvertent contact with wire baods)					IP 44		
5 Protection against dust (Harmful deposite of dust)					IP 54	IP 55	IP 56

Fig. 92 IP CLASSIFICATIONS OF MOTORS (Source IS- 4691)

FIRST CHARACTERISTIC NUMERAL \ SECOND CHARACTERISTIC NUMERAL		Free Convection 0	Self-Circulation 1	Integral Component Mounted on Separate Shaft 2	Dependent Component Mounted on the Machine 3	Integral Independent Component 5	Independent Component Mounted on the Machine 6	Independent and Separate Device or Coolant System Pressure 7	Relative Displacement 8
Free circulation	0	IC 00	IC 01 or IC 0						
Inlet pipe ventilated	1		IC 11 or IC 1						
Outlet pipe ventilated	2		IC 21 or IC 2						
Inlet and outlet pipe ventilated	3		IC 31 or IC 3						
Frame-surface cooled	4		IC 41 or IC 4						
Integral heat exchange (using surrounding medium)	5		IC 51 or IC 5						
Machine mounted heat exchanger (using surrounding medium)	6		IC 61 or IC 6						
Integral heat exchanger (not using surrounding medium)	7								
Machine mounted heat exchanger (not using surrounding medium)	8								
Separately mounted heat exchanger	9								

Fig. 93. COOLING CLASSIFICATION (IC) OF MOTORS (Source:ISS 6362)

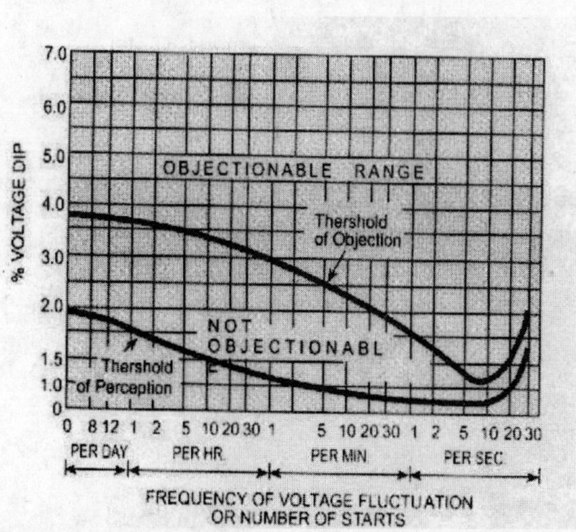

Fig. 94. VOLTAGE FLICKER CURVE (Source: IEEE standard-141-1993)

Table 95: INDUCTION MOTOR QUICK SELECTION CHART

Classification	Starting torque (% of FLT)	Breakdown torque (% of FLT)	Starting current	Slip	Typical applications
Normal starting torque and normal starting current	100-200	200-500	Normal	< 5%	Fans, blowers, centrifugal pumps, compressors, etc.
High starting torque and normal starting current	200-250	200-250	Normal	< 5%	Conveyors, crushers, stirring machines, agitators, reciprocating pumps and compressors, etc.
High starting torque and high slip	275	275	low	> 5%	High peak loads, load with flywheels like punch wheels, shears, elevators, extractors, winches, hoists, wire drawing machines etc.
Wound rotor	Any torque up to breakdown torque	225-275	Depends on starting torque	Depends on rotor resistance	Where high starting torque with low inrush current, frequent start, or limited to 2:1 speed control are required, and where high inertia must be accelerated

Table 96: SELECTION OF MOTOR CAPACITY FOR INTERMITTENT LOADING

When the motors are subjected to varying loading conditions for varying periods of time, computation of the rating of the motor will be done using the equation:

$$kW_{rms} = \frac{\sqrt{\sum_{1}^{n}\left(kW_i^2 \times t_i\right)}}{\sqrt{\left(\frac{t_0}{k_0} + \sum_{1}^{n} t_i\right)}}$$

where,

kW_{rms} = rms rating of the motor in kilowatts

kW_i = loading of the motor at the ith instant

t_i = time duration of the ith instant

k_0 = is a constant to take care of the rest period

= 3 for open type motors

= 4 for totally enclosed type motors

t_0 = time period of rest for the motor

Table 97: BASIC FUNCTION OF SWITCHGEARS

Electrical protection against	Isolation	Control
Overload current Short circuits Insulation failure	Isolation clearly indicated by a fail proof mechanical indicator. A gap interposed insulating barrier between the open contacts, clearly visible.	Functional switching Emergency switching Emergency stopping Switching off for maintenance

Table 98: UTILIZATION CATEGORIES OF LOAD BREAK SWITCHES (IEC-60947-3)

Utilization category		Typical applications	cos φ	Making current	Breaking current
Frequent operation	**Infrequent operation**				
AC-20A	**AC-20B**	Connecting and disconnecting under no load conditions	-	-	-
AC-21A	**AC-21B**	Switching of resistive loads with moderate overloads	0.95	$1.5I_n$	$1.5I_n$
AC-22A	**AC-22B**	Switching of mixed resistive and inductive loads with moderate overloads	0.65	$3I_n$	$3I_n$
AC-23A	**AC-23B**	Switching of motor loads or other highly inductive loads	0.45 for I < 100A 0.35 for I > 100A	$10I_n$	$8I_n$

Table 99: TYPICAL SPECIFICATIONS FOR LOAD BREAK SWITCHES

Description	Unit	Ratings		
Rated current at 40ºC	A	125	160	200
Rated insulation voltage	V	1000	1000	1000
Rated operational voltage	V	415	415	415
Dielectric strength 50 Hz	kV	5	5	5
Rated impulse withstand voltage	kV	10	10	10
Rated operational current at 415 V AC-21A/B AC-22A/B		 125/125 125/125	 125/125 125/125	 125/125 125/125

AC-23A/B		125/125	125/125	125/125
Rated making capacity-AC-23A at 0.3 pf	A	1250	1600	1600
Rated breaking capacity-AC-23A at 0.3 pf	A	1000	1280	1280
Rated motor power	kW	55	55	100
Fuse operated short circuit withstand				
Rated conditional short circuit current	kA_{rms}	80	80	80
Max allowed cut off current	kA_{peak}	17	18	22
Rated short time withstand current (1 sec)	kA_{rms}	7.5	7.5	7.5
Electrical durability-operating cycles-AC-23A/B		3000/600	3000/600	3000/600
Mechanical durability-operating cycles		30000	30000	30000
Temperature range	ºC	-5 to 50	-5 to 50	-5 to 50

Table 100: TYPICAL SPECIFICATIONS FOR CHANGE-OVER SWITCHES

Current rating at 40 ˚C	**A**	**125**	**160**	**200**
Number of poles		4	4	4
Rated insulation voltage	V	1000	1000	1000
Rated operational voltage	V	415	415	415
Rated impulse withstand voltage	kV	10	10	10
Rated operational current at 415 V				
AC-21A/B	A	125/125	160/160	200/200
AC-22A/B	A	125/125	160/160	200/200
AC-23A/B	A	125/125	160/160	200/200
Rated making capacity	A	1250	1600	1600
Rated breaking capacity	A	1000	1280	1280
Fuse protected short circuit withstand				
Rated max current of gG fuse	A	125	160	200
Rated conditional short circuit current	kA	80	80	80
Max allowed cut off current	kA	17	18	22
Electrical durability No. of operating cycles-AC-23A/B		3000/600	3000/600	3000/600
Mechanical durability No. of no load operating cycles		30000	30000	30000
Terminal connection cable size (Al)	mm^2	70	95	150
Temperature withstand range	˚C	-5 to 50	-5 to 50	-5 to 50

(Source: Havell's India Ltd.)

Table 101: UTILISATION CATEGORIES OF AUTOMATIC TRANSFER SWITCHES (IEC 947-6-1)

Supply	Utilization category		Applications
	Frequent operation	**Infrequent operation**	
AC	AC-31A AC-33A AC-35A AC-36A	AC-31B AC-33B AC-35B AC-36B	Non-inductive or slightly ind. loads Motor loads or mixed loads Electric discharge lamps Incandescent lamps.
DC	DC-31A DC-33A DC-36A	DC-31B DC-33B DC-36B	Resistive load, Motor loads or mixed loads, Incandescent lamp load.

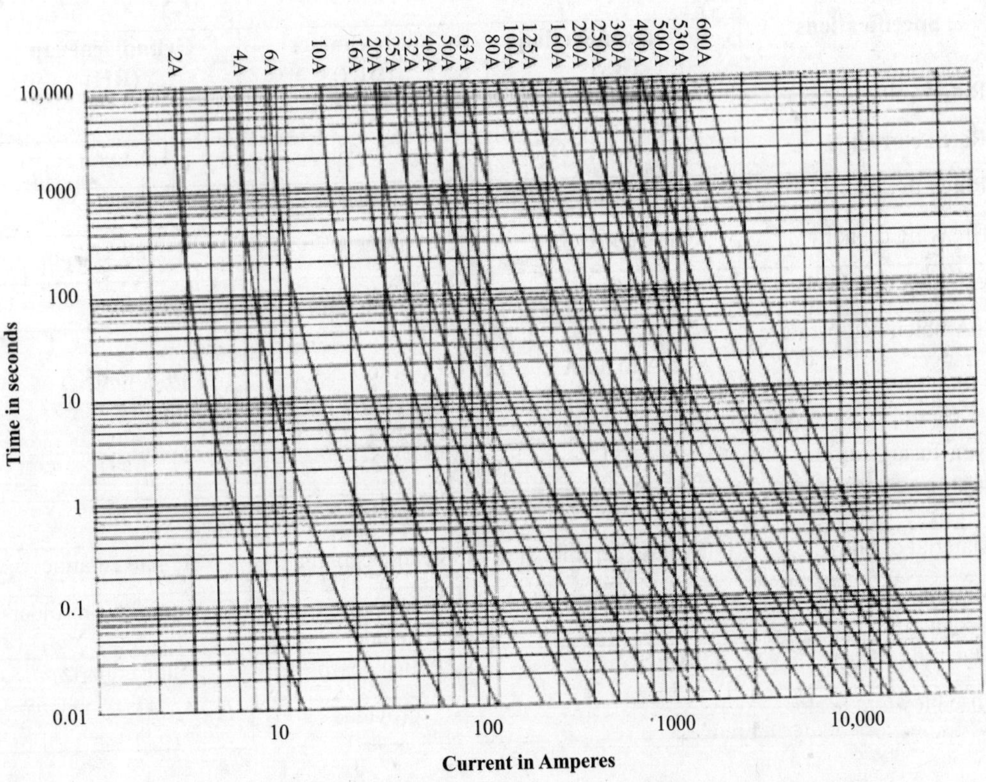

Fig. 102: TIME-CURRENT CHARACTERISTICS OF FUSES

Table 103: FUSING AND NON-FUSING ZONES FOR MEDIUM VOLTAGE gG AND gM FUSES (IEC 60269)

Rated current I_n	Non-fusing current I_{nf}	Fusing current I_f	Conventional time (hours)
$I_n \leq 4A$	$1.5\ I_n$	$2.1\ I_n$	1
$4 < I_n \leq 16\ A$	$1.5\ I_n$	$1.9\ I_n$	1
$16 < I_n \leq 63\ A$	$1.25\ I_n$	$1.6\ I_n$	1
$63 < I_n \leq 160\ A$	$1.25\ I_n$	$1.6\ I_n$	2
$160 < I_n \leq 400\ A$	$1.25\ I_n$	$1.6\ I_n$	3
$400 < I_n$	$1.25\ I_n$	$1.6\ I_n$	4

Table 104: TYPICAL SPECIFICATIONS OF TYPE gG FUSES (Source: Havells India Ltd.)

Specifications	Type of fuse		
	Bolted connection (BS)	Blade contact (DIN)	Cylindrical cap (RH)
Rated voltage	415 V	415 V	415 V
Rated current	2 A to 630 A	6 A to 630 A	4 A to 63 A
Rated frequency	50 Hz	50 Hz	50 Hz
Breaking capacity	80 kA	80 kA	80 kA
Utilization category	gG	gG	gG
Ambient temperature range: 70°C 65°C 60°C	2 A to 32 A 35 A to 80 A	6 A to 32 A 35 A to 80 A 100 A to 630 A	4 A to 63 A
Non-fusing current	$1.25I_n$	$1.25I_n$	$1.25I_n$
Fusing current	$1.6I_n$	$1.6I_n$	$1.6I_n$
Material of body	Steatite ceramic	Steatite ceramic	Steatite ceramic
Material of fusing element	Copper with tin solder	Copper with tin solder	Copper with tin solder
Material of filler	Silica quartz	Silica quartz	Silica quartz
Fuse blown indicator	-	provided	-

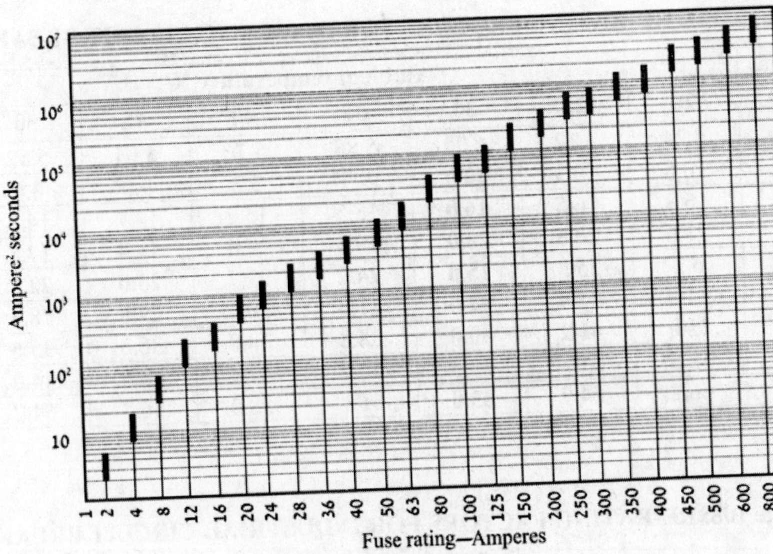

Fig. 105. FUSELET THROUGH ENERGY VS CURRENT RATING
(Source: Havells India)

Table 106: TRIPPING CURRENT RANGES OF OVERLOAD AND SHORT CIRCUIT PROTECTIVE DEVICES

Type of device	Type of protective relay	Overload protection	Short circuit protection		
Domestic breakers-IEC 60898	Thermal magnetic	$I_r = I_n$	Low setting type B $3I_n \le I_m \le 5I_n$	Standard setting type C $5I_n \le I_m \le 10I_n$	High setting type D $10I_n \le I_m \le 20I_n$
Modular industrial breakers	Thermal magnetic	$I_r = I_n$ fixed	Low setting type B or Z $3.2I_n \le \text{fixed} \le 4.8I_n$	Standard setting type C $7I_n \le \text{fixed} \le 10I_n$	High setting type D or K $10I_n \le \text{fixed} \le 14I_n$
Industrial circuit breakers IEC 60947-2	Thermal magnetic	$I_r = I_n$ fixed	Fixed, $I_m = 7$ to $10\ I_n$		
		Adjustable $0.7I_n \le I_r \le I_n$	Adjustable Low setting : 2 to 5 I_n Standard setting : 5 to 10 I_n		
	Electronic	Long delay $0.4I_n \le I_r \le I_n$	Short delay – adjustable $1.5I_r \le I_m \le 10\ I_r$ Instantaneous(I)fixed I =12 to 15 I_n		

Table 107: RE-RATING FACTORS FOR DOMESTIC CIRCUIT BREAKERS

Current rating(A)	Ambient temperature °C							
	20	**25**	**30**	**35**	**40**	**45**	**50**	**55**
6	6.24	6.12	**6.00**	5.88	5.76	5.64	5.52	5.4
10	10.6	10.3	**10.0**	9.7	9.3	9.0	8.6	8.2
16	16.8	16.5	**16.0**	15.5	15.2	14.7	14.2	13.8
20	21.0	20.6	**20.0**	19.4	`9.0	18.4	17.8	17.4
25	6.2	25.7	**25.0**	24.2	23.7	23.0	22.2	21.5
32	33.5	32.9	**32.0**	31.4	30.4	29.8	28.4	28.2
40	42.0	41.2	**40.0**	38.8	38.0	36.8	35.6	34.4
50	52.5	51.5	**50.0**	48.5	47.4	45.5	44.0	42.5
63	66.2	64.9	**63.0**	61.1	58.0	56.7	54.2	51.7

Table 108: RE-RATING FACTORS FOR INDUSTRIAL CIRCUIT BREAKERS
(Source: Schneider Electric)

Current rating(A)	Ambient temperature °C							
	20	**25**	**30**	**35**	**40**	**45**	**50**	**55**
16	18	18	17	16.6	**16**	15.6	15.2	14
25	27.5	25	26.3	25.6	**25**	24.5	24	23.5
32	35.2	34.4	33.6	32.8	**32**	31.3	30.5	30
40	44	43	42	41	**40**	39	38	37
50	55	54	52.5	51	**50**	49	48	47
63	69	68	66	65	**63**	61.5	60	58
80	88	86	84	82	**80**	78	76	74
100	110	108	105	103	**100**	97.5	95	92.5
125	138	134	131	128	**125**	122	119	116
160	176	172	168	164	**160**	156	152	148
200	220	215	210	205	**200**	195	190	185
250	277	269	263	256	**250**	244	238	231

Table 109: TYPICAL DATA OF MOULDED CASE CIRCUIT BREAKERS (MCCB)
(Source: Legrand (India) Pvt. Ltd.)

Type and frame size		**DPX 125**			**DPX 160**		**DPX 250 ER**		**DPX 250**			**DPX 250 electronic release**	**DPX 630**		
Number of poles		3-4	3-4	3-4	3-4	3-4	3-4		3-4			3-4	3-4		
Normal rating in (A)		16-125	16-125	16-125	160		100-250		25-250			40-250	400-630		
Electrical characteristic conforming to EN 60947-2 and IS 13947-2															
Rated operating voltage U_e (V dc)		500			500		500		690			690	690		
Rated insulating voltage U_i (V ac)		250			250		250		250			-	250		
Rated impulse withstand voltage		500			500		500		690			690	690		
Rated impulse withstand voltage U_{imp} (kV)		6			6		6		8			8	8		
Category of use		A			A		A		A			A	A		
Ultimate breaking capacity I_{cu}(kA)	400/415VA	16	25	36	36	50	36	50	36	70	100	70	36	70	100
Standard braking capacity I_{cs} (% I_{cu})		100	50	75	75	50	75	50	100	75	50	75	100	75	50
Rated closing capacity on short circuit		32	52.5	75.6	75.6	100	75.6	105	75.6	154	220	154	75.6	154	220
Endurance (o.c cycle)	mechanical	25000			20000		20000		20000			20000	15000		
	electrical	8000			8000		8000		8000			8000	5000		

Table 110: TYPICAL DATA OF AIR CIRCUIT BREAKERS (SOURCE: SCHNEIDER ELECTRIC CO)

Common Characteristics													
Number of poles		3/4											
Rated insulation voltage (v)	**Ui**	1000/1250											
Impulse withstand voltage (kv)	**Uimp**	12											
Rated operational voltage (v AC 50\60 Hz)	**Ue**	690/1150											
Suitability for isolation	IEC 60947-2												
Degree of pollution	IEC 60664-1	4(1000V)/3 (1250 V)											
Circuit breaker characteristics as per IEC 60947-2			**NWO8**	**NW10**	**NW12**	**NW16**	**NW20**	**NW25**	**NW32**	**NW40**	**NW40b**	**NW50**	**NW63**
Rated current (A)		at 40°/50°C	800	1000	1250	1600	2000	2500	3200	4000	4000	5000	6300
Rating of 4th pole (A)			800	1000	1250	1600	2000	2500	3200	4000	4000	5000	6300
Sensor rating (A)			400 to 800	400 to 1000	630 to 1250	800 to 1600	1000 to 2000	1250 to 2500	1600 to 3200	2000 to 4000	2000 to 4000	2500 to 5000	3200 to 6300
Type of circuit breaker			**N1**	**H1**	**H2**	**L1(2)** **H10**	**H1 H2 H3** **LI(2) H10**	**H1**	**H2**	**H3** **H10**	**H1**	**H2**	
Ultimate breaking capacity (kA rms) V AC 50/60Hz	I_{cu}	220/415/ 440V	42	65	100	150 -	65 100 150 150 -	65	100	150 -	100	150	
Rated service breaking capacity (kA rms)	I_{cs}	$\%I_{cu}$	100% B				100% B	100% B			100% B		
Utilisation category Rated short-time withstand current (kA rms)	I_{cw}	1s	42	65	85	30/50	65 85 65 30 50	65	85	65	50	100	100
V AC 50/60Hz		3s	22	36	50	30/50	36 75 65 30 50	65	75	65	50	100	100
Integrated instantaneous protection (kA peak±10%)			-	-	190	80 -	- 190 150 80 -	-	190	150 -	+	270	
Rated making capacity (kA peak)	I_{cm}	220/415/ 440V	88	143	220	330	143 220 330 330 -	143	220	330 -	220	330	
V AC 50/60 Hz													
Break time (ms) between tripping order and arc extinction			25	25	25	10 25	25 25 25 10 25	25	25	25 25	25	25	
Closing time (ms)			<70				<70	<70			<80		

Table 110: Contd.

Circuit breaker characteristics as per NEMA AB1									
Breaking capacity (kA)		240/480V	42 65 100 150 - 65 150 100 150 - 65 100 150 - 100 150						
Unprotected circuit breaker characteristics:									
Tripping by shunt trip as per IEC 60947-2									
Type of circuit breaker			**HA**	**HF(3)**	**HA**	**HF(3)**	**HA**	**HF(3)**	**HA**
Ultimate breaking capacity (kA rms)V AC 50/60 Hz	I_{cu}	220..690V	50	85	50	85	55	85	85
Rated service breaking capacity (kA rms)	I_{cs}	$\% I_{cu}$	100%		100%		100%		100%
Rated short-time withstand current (kA rms)	I_{cw}	1 s	50	85	50	85	55	85	85
		3s	36	50	36	75	55	75	85
short circuit protection, maximum delay: 350 ms (4)									
Rated making capacity (kA peak) VAC 50/60 Hz	I_{cm}	220..690V	105	187	105	187	121	187	187

Table 111: TYPES OF SYSTEM EARTHING

Wiring regulations use a series of letter designations to precisely define the relationship between the source of supply (usually, the secondary of distribution transformers) including the neutral and of the exposed conductive parts and earth.

The first letter denotes the earthing arrangement of the source of energy. They are:

T = Direct connection of one or more points to earth (**T**erra)

I = All active parts isolated from earth or one point connected to earth through an impedance.

The second letter denotes the relationship of the exposed conductive parts of the installation to earth. These letters are:

T = Direct electrical connection of the exposed conductive parts to earth, independently of the earthing of any point of the source of energy.

N = Direct electrical connection of the exposed conductive parts to the earthed point of the source of energy, which for AC systems is the **N**eutral point of the supply transformer.

The remaining letters describe the arrangement of the neutral conductor and the protective conductor.

S = Conductor which is **S**eparated for the neutral or the earthed conductor. (In DC system this is an earthed negative or positive conductor

C = Neutral conductor and protective functions **C**ombined in a single conductor (PEN conductor)

PE = **P**rotective **E**arth conductor

N = **N**eutral conductor

According to the wiring regulations, the three main distribution systems are:

- IT system
- TN system
- TT system

These systems of earthing are shown below:

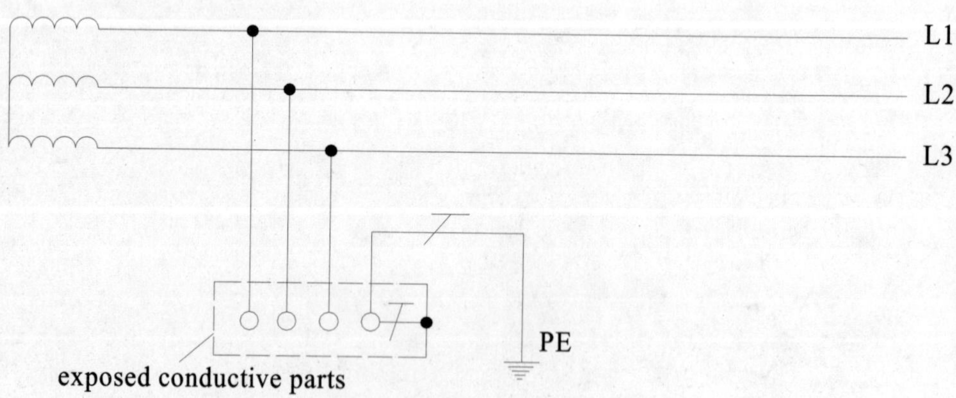

IT system of earthing

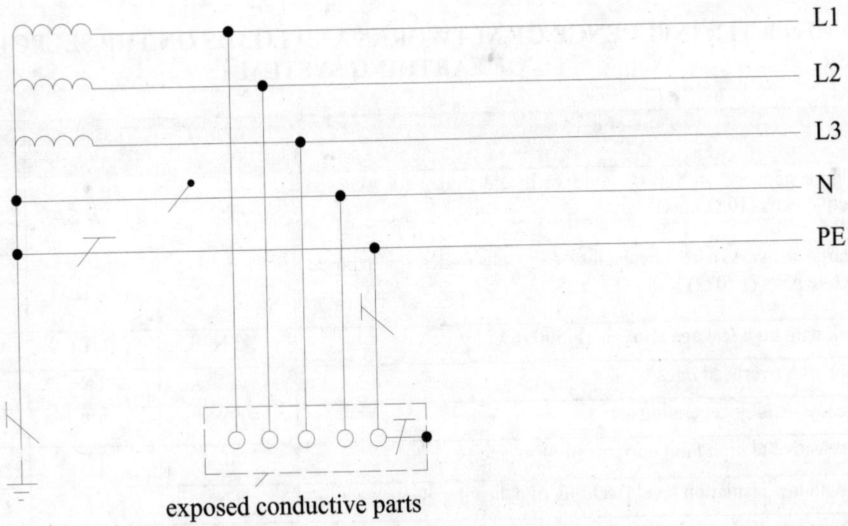

TN system of earthing

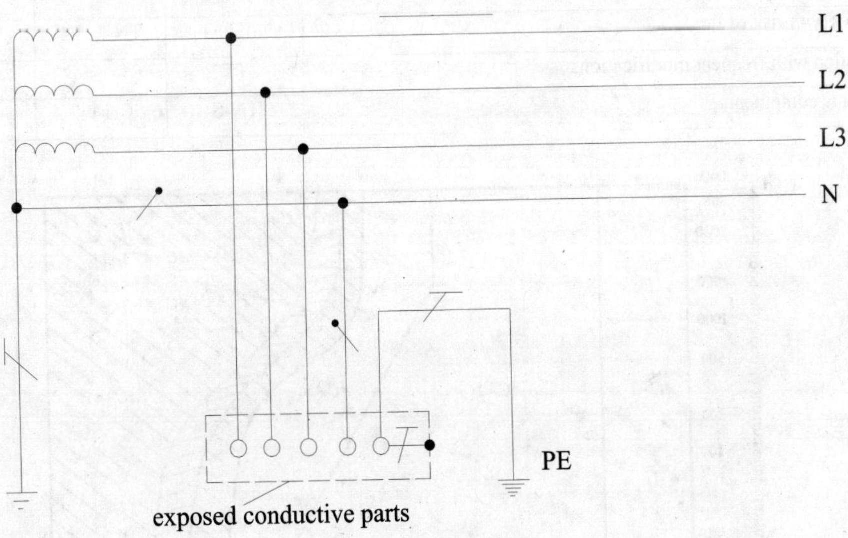

TT system of earthing

Table 112: INFLUENCE OF NETWORKS AND LOADS ON THE SELECTION OF EARTHING SYSTEM

Type of network	Advised	Possible	Not advised
Very large network with very good earth electrodes for exposed conductive parts (10 Ω max)	TN	TN-S	IT, TN-C
Very large network with low quality earth electrodes for exposed conductive parts (>30 Ω)	TN	TT	IT
Network with high leakage currents (> 500 mA)	TN	IT,TT	
Network with overhead lines	TT	TN	IT
Emergency standby generating set	IT	TT	TN
Loads sensitive to high fault currents (motors, etc.)	It	TT	TN
Loads with low insulation level (welding m/c, heating elements, etc)	TN	TT	IT
Single phase portable loads	TT, TN-S		IT, TN-C
Loads with sizable risks (conveyors, hoists, etc)	TN	TT	IT
Machine tools	TN-S	TN-C, IT	TT
Supply via star connected power transformer	TT	IT w/o neutral	IT with neutral
Premises with risk of fire	IT	TN-S, TT	TN-C
Installation with frequent modifications	TT		TN, IT
Electronic equipment	TN-S	TT	TN-C

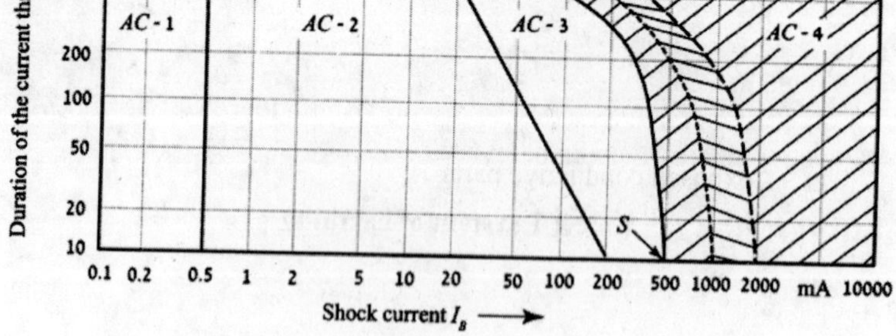

Fig. 113: EFFECT OF ALTERNATING CURRENT ON HUMAN BODY (IEC Standard 60479)

Table 113: EFFECT OF ALTERNATING CURRENT ON HUMAN BODY (IEC Standard 60479)

Current zones	Shock current	Duration of current flow	Physiological effects
AC-1	0-1 mA	Uncritical	Usually no reaction—no electrification or only slight electrification perceptible
AC-2	1-10mA	Uncritical	Strong, partly painful effects on muscles, fingers, feet and joints of arms and legs
AC-3	10-500 mA	Minutes up to split seconds	Strong and painful effects on arms, legs and shoulder muscles, rise in blood pressure, difficulty in breathing. Reversible interferences in the heart, unconsciousness in the upper range.
AC-4-1	Between C_1 and C_2		Ventricular Fibrillation unlikely to occur (probability blow 5%)
AC-4-2	Between C_2 and C_3		Possible ventricular fibrillation (probability up to 50 %)
AC-4-3	Via C_3		High risk of ventricular fibrillation (above 50%)

Table 114: POWER FACTOR

Definition

The power factor PF is the ratio between the active power and the apparent power

$$PF = \frac{P}{S}$$

There is always confusion among the practicing engineers about the value of cos φ and the power factor PF. The value of cos φ is given by

$$\cos\varphi = \frac{P_1}{S_1}$$

where

P = Active power
S = Apparent power
P_1=Active power of the fundamental component of the frequency
S_1= Apparent power of the fundamental frequency

Table 115: POWER QUALITY ISSUES

The total harmonic distortion, THD of the voltage is given by

$$THD_V = \frac{\sqrt{V_2^2 + V_3^2 + ... + V_n^2}}{V_1}$$

$$= \frac{\sqrt{\sum_{h=2}^{n} V_h^2}}{V_1}$$

Similarly, THD of current waveforms can also be given as:

$$THD_I = \frac{\sqrt{\sum_{h=2}^{n} I_h^2}}{I_1}$$

And power factor, $$PF = \frac{\cos\varphi_1}{\sqrt{1 + THD_i^{\,2}}}$$

Table 116: RELATION BETWEEN THD_I AND PF

THD_I	0%	20%	40%	60%	80%	100%
$\frac{PF}{\cos\varphi}$	1.00	0.981	0.9285	0.8572	0.781	0.707

Table 117: HARMONIC CURRENT SPECTRUM OF THREE-PHASE VARIABLE SPEED DRIVE AND THREE-PHASE FULL WAVE RECTIFIERS

Harmonic order	Amplitude (% of Fundamental)	
	6 pulse	12 pulse
1	100.00	100.00
5	20.00	
7	14.30	
11	9.10	9.10
13	7.70	7.70
17	5.90	
19	5.30	
23	4.30	4.30
25	4.00	4.00

Table 118: CURRENT DISTORTION LIMITS FOR GENERAL DISTRIBUTION SYSTEMS (120 V through 69000 V) Source: IEEE Standard 519-1992

Maximum Harmonic current distortion in % of I_L						
Individual Harmonic order (Odd Harmonics)						
$\frac{I_{SC}}{I_L}$	< 11	11 ≤ h < 17	17 ≤ h < 23	23 ≤ h < 35	35 ≤ h	THD_I (TDD)
< 20*	4.0	2.0	1.5	0.6	0.3	5.0
20 < 50	7.0	3.5	2.5	1.0	0.5	8.0
50 < 100	10.0	4.5	4.0	1.5	0.7	12.0
100 < 1000	12.0	5.5	5.0	2.0	1.0	15.0
> 1000	15.0	7.0	6.0	2.5	1.4	20.0

Even harmonics are limited to 25% of the odd harmonic limits. TDD refers to total demand distortion and is based on the average maximum demand current at the fundamental frequency measured at point of common coupling (PCC).

* All power generation equipment is limited to these values of current distortion regardless of short circuit current ratio.

I_{SC} – Maximum short circuit current at PCC
I_L – Maximum demand load current at PCC
H – Harmonic order

Table 119: RECOMMENDED VOLTAGE DISTORTION LEVELS (Source: IEEE Stamdard 519-1992)

Bus voltage at PCC	Individual voltage distortion %	Total Harmonic Distortion THD_V %
69 kV and below	3.0	5.0
69.001 through 161 kV	1.5	2.5
161.001 and above	1.0	1.5

These limits should be used as system design values for the **worst case** under normal operation (lasting longer than one hour). For shorter periods like starting-up or unusual conditions the limits may exceed by 50 %.

Table 120: DETUNED HARMONIC FILTER DESIGN

The basic problem involved in the design of detuned harmonic filter design can be summarized as follows.

Step-1 Determine the reactive kVA compensation required.

Step-2 Select a capacitor bank with a voltage rating slightly above the normal rated voltage of the supply system or select a low stress type capacitor.

Step-3 Determine the de-rated capacity of the capacitor bank using equation.

Step-4 Calculate the rating of the tuning inductor to tune the capacitor to the desired harmonic frequency.
Step-5 Measure/calculate the harmonic spectrum of the load current.
Step-6 Calculate the harmonic spectrum of the voltages and the effective rms bus voltage.
Step-7 Calculate the harmonic spectrum of the source current and the effective rms source current.
Step-8 Check for consistency with IEEE 519 standard.

It is to be noted that IEEE 519 1992 is meant to apply to harmonic distortion of the system at PCC level rather than to the harmonic distortion levels of the individual loads. Therefore, any attempt to apply these limits to individual loads is not in good spirit and such an interpretation will only involve additional expenses and should be avoided as far as possible.

Table 121: ROOF AREA REQUIREMENT OF SOLAR PV SYSTEMS

PV Module Efficiency (%)	PV Capacity (kilowatt)							
	0.1	0.25	0.5	1	2	4	10	100
	Approximate Roof area needed in sq m							
4	3	7.5	15	30	60	120	300	3000
8	1.5	3.8	7.5	15	30	60	150	1500
12	1	2.5	5	10	20	40	100	1000
16	0.8	2	4	8	16	32	80	800

Table 122: TECHNICAL SPECIFICATIONS OF SOLAR PV MODULES (Courtesy: M/s Indsolar Pvt Ltd)

Power (Pm) in watts (Nominal)	Open circuit voltage (Voc) in volts	Short circuit current (Isc) in Amps	Voltage at maximum power (Vmp) in volts	Current at maximum power (Imp) in Amps	Area sq.m/ (Area sq.ft)
40	21	2.54	17	2.35	0.35(3.76)
50	21	3.17	17	2.94	0.43(4.63)
60	21	3.81	17	3.53	0.50(5.38)
70	21	4.44	17	4.12	0.56(6.03)
74	21	4.7	17	4.35	0.61(6.66)
75	21	4.75	17	4.41	0.61(6.66)
80	21	5.08	17	4.71	0.62(6.67)
90	21	5.71	17	5.29	0.70(7.53)
100	21	6.35	17	5.88	0.78(8.40)
110	21	6.98	17	6.47	0.85(9.25)

120	21	7.62	17	7.06	1.0009(10.86)
125	21	7.94	17	7.35	
130	21	8.25	17	7.65	
140	21	8.89	17	8.24	
150	43	4.65	35	4.29	1.163(12.52)
160	43	4.96	35	4.57	

Table 123: SPECIFICATIONS OF SOLAR PV MODULES

Type	Polycrystalline silicon
Origin	Manufactured in India
Efficiency	>=13%
Fill factor	>=70%
Degradation warranty	Panel output (Wp) capacity to be >=90% of design nominal power after 10 years and >= 80% of design nominal power after 25 years.
Module frame	Non-corrosive and electrolytically compatible with the mounting structure material.
Termination box	Thermoplastic, IP 65, UV-resistant
Blocking diodes	Schottky type
Module minimum rated power	The nominal power of a single PV module shall not be less than 74 Wp.
RF identification tag for each solar module	Shall be provided inside the module and must be able to withstand environmental conditions and last the lifetime of the solar module.
RF identification tag data	(a) Name of the manufacturer of PV module
	(b) Name of manufacturer of solar cells
	(c) Month and year of manufacture (separately for solar cells and module)
	(d) Country of origin (separately for solar cells and module)
	(e) I-V curve for the modules
	(f) W_m, I_m, V_m and FF for the module
	(g) Unique serial no. and model no. of the module
	(h) Date and year of obtaining IEC PV module qualification certificate
	(i) Name of the test lab issuing IEC certificate
	(j) Other relevant information on traceability of solar cells and module as per ISO 9000 standard
Power output rating	To be given for standard test conditions (STC). I-V curve of the sample module shall be submitted.

Compliance with standards and codes	IEC 61215/IS 14286 IEC 61730 Part 1 and 2
Salt Mist Corrosion Testing	As per IEC 61701

Table 124: SOLAR PV MODULES MOUNTING STRUCTURE (Courtsey: M/s Indsolar Pvt Ltd)

Wind velocity with standing capacity	150 km/hour
Structure material	Hot dip galvanized steel with a minimum galvanization thickness of 120 microns or aluminium alloy.
Bolts, nuts, fasteners, panel mounting clamps	Stainless steel SS 304
Mounting arrangement for RCC-flat roofs	With removable concrete ballast made of pre-fabricated PCC (1:2:4), M15
Mounting arrangement for metal sheet roofs	Mounting directly on the sheet metal ensuring stability and wind withstanding capacity, or penetrating the sheet metal and fixing to the sub-structure, ensuring that the roof remains water-proof and ensuring stability and wind withstanding capacity.
Mounting arrangement for elevated structures	The elevated structure has to be securely anchored to the supporting surface. Concrete foundations of appropriate weight and depth for elevated structures mounted directly on the ground; bolted with anchor bolts of appropriate strength for elevated structures mounted on RCC surfaces.
Mounting arrangement for ground installations	With removable concrete ballast made of pre-fabricated PCC (1:2:4), M15; assuring enough ground clearance to prevent damage of the module through water, animals and other environmental factors.
Installation	The structures shall be designed for simple mechanical on-site installation. There shall be no requirement of welding or complex machinery at the installation site.
Minimum distance between roof edge and mounting structure	0.6 m
Access for panel cleaning and maintenance	All solar panels must be accessible from the top for cleaning and from the bottom for access to the module junction box.
Panel tilt angle	North-south orientation with a fixed tilt angle of 11-13 degrees (depending on location), south facing.

Table 125: CHARGING CHARACTERISTICS OF SOLAR BATTERIES

12 V Low Maintenance Tubular Floated- LMS Series (Courtesy: Exide Batteries)

Model of operation	Voltage setting per mono-block unit for ambient- temperature		Current setting
	12 V mono-block	6 V mono-block	
Bulk voltage	15.6 V ± 0.1 V	7.80 V ± 0.1 V	Maximum-20% of battery Ah capacity Minimum-10% of battery Ah capacity
Float voltage	13.8 ± 0.1 V	6.90 V ± 0.1 V	
Low voltage disconnect	11.4 V ± 0.1 V	5.70 V ± 0.1 V	
Load reconnect voltage	12.96 V ± 0.1 V	6.48 V ± 0.1 V	
Recharge factor	110% of discharge Ah		

Technical data

Type of battery	Nominal voltage (V)	Capacity @ 10 hr to 1.80 V at 27°C (Ah)
6 LMS 20	12	20
6 LMS 20 L	12	20
6 LMS 40	12	40
6 LMS 40 L	12	40
6 LMS 75	12	75
6 LMS 75 L	12	75
6 LMS 100 L	12	100
6 LMS 120 L	12	120
6 LMS 150 L	12	150
6 LMS 200 L	12	200
3 LMS 300	6	300

Table 126: TUBULAR GEL VRLA BATTERIES RECHARGING CHARACTERISTICS DURING OPERATIONS (Courtesy: Exide Batteries)

	Recommended parameters for ambient temperature of 25-30°C	Temperature compensation factor at 25°C
Charging current	Maximum-20% of battery Ah capacity Minimum-10% of battery Ah capacity	
Bulk voltage	2.40 ± 0.02 V/cell	Float : -3 mV / °C / 2 V cell
Float voltage	2.28 ± 0.02 V/cell	
Equalizing voltage	2.40 ± 0.02 V/cell	
Load reconnect voltage	2.20 ± 0.02 V/cell	
Low voltage disconnect	1.90 ± 0.02 V/cell	Cyclic : -5 mV/°C/2 V cell
Recharge factor	105% of discharge Ah	

Technical Data

Type of battery	Nominal voltage (V)	Capacity @ 10 hr / 1.75 V (Ah)
SG 200	2 V	200
SG 300	2 V	300
SG 400	2 V	400
SG 500	2 V	500
SG 600	2 V	600
SG 800	2 V	800
SG 1000	2 V	1000

Table 127: VALVE REGULATED LEAD ACID BATTERIES (VRLA) Discharge Characteristics in Watts for Long Duration Discharges

Module type	Nominal capacity (Ah) at C_{10}	Discharge in hours (1.75 ECV)						Discharge in hours (1.80 ECV)					
		20	24	48	72	100	120	20	24	48	72	100	120
T-80 HP	80	4.4	3.8	2.2	1.5	1.1	1	4.3	3.7	2.1	1.5	1.1	1
T-100 HP	100	5.5	4.8	2.7	1.9	1.4	1.3	5.4	4.6	2.6	1.8	1.3	1.2
T-120 HP	120	6.6	5.7	3.3	2.3	1.7	1.5	6.4	5.5	3.1	2.2	1.6	1.5
T-160 HP	160	8.8	7.7	4.3	3	2.2	2	8.6	7.3	4.2	2.9	2.2	1.9
T-200 HP	200	11	9.6	5.4	3.8	2.8	2.6	10.7	9.2	5.2	3.6	2.7	2.4
T-240 HP	240	13.2	11.5	6.5	4.5	3.4	3.1	12.9	11	6.3	4.4	3.2	2.9
T-280 HP	280	15.4	13.4	7.6	5.3	3.9	3.6	15	12.8	7.3	5.1	3.8	3.4
T-300 HP	300	16.5	14.4	8.1	5.7	4.2	3.8	16.1	13.8	7.8	5.5	4	3.6

T-320 HP	320	17.6	15.3	8.7	6.1	4.5	4.1	17.2	14.7	8.3	5.8	4.3	3.9
T-360 HP	360	19.8	17.2	9.8	6.8	5	4.6	19.3	16.5	9.4	6.6	4.8	4.4
T-400 HP	400	22	19.2	10.9	7.6	5.6	5.1	21.4	18.3	10.4	7.3	5.4	4.9
T-440 HP	440	24.2	21.1	11.9	8.3	6.2	5.6	23.6	20.2	11.5	8	5.9	5.3
T-480 HP	480	26.4	23	13	9.1	6.7	6.1	25.7	22	12.5	8.7	6.5	5.8
T-500 HP	500	27.5	24	13.6	9.5	7	6.4	26.8	22.9	13	9.1	6.7	6.1
T-520 HP	520	28.6	24.9	14.1	9.8	7.3	6.6	27.9	23.8	13.5	9.5	7	6.3
T-560 HP	560	30.8	26.8	15.2	10.6	7.8	7.1	30	25.7	14.6	10.2	7.5	6.8
T-600 HP	600	33	28.7	16.3	11.4	8.4	7.7	32.2	27.5	15.6	10.9	8.1	7.3
T-650 HP	650	35.8	31.1	17.6	12.3	9.1	8.3	34.9	29.8	16.9	11.8	8.7	7.9
T-680 HP	680	37.4	32.6	18.5	12.9	9.5	8.7	36.5	31.2	17.7	12.4	9.2	8.3
T-760 HP	760	41.8	36.4	20.6	14.4	10.6	9.7	40.8	34.8	19.8	13.8	10.2	9.2
T-850 HP	850	46.8	40.7	23.1	16.1	11.9	10.8	45.6	39	22.1	15.5	11.4	10.3
T-925 HP	925	50.9	44.3	25.1	17.5	12.9	11.8	49.6	42.4	24.1	16.8	12.5	11.2
T-1000 HP	1000	55	47.9	27.1	18.9	14	12.8	53.6	45.9	26.1	18.2	13.5	12.1

Table 128: VALVE REGULATED LEAD ACID BATTERIES Discharge Characteristics in Watts 1.75 ECV

Module type	Nominal capacity (Ah) at C_{10}	Discharge in minutes				Dischrge in hours										
		5	10	15	30	1	1.5	2	3	4	5	6	7	8	9	10
T-80	80	220	179	156	119	87	66	54	41	34	26	23	20	19	17	16
T-100	100	275	224	194	149	108	83	68	51	43	33	28	25	23	21	19
T-120	120	330	269	233	179	130	99	81	61	52	40	34	31	28	25	23
T-160	160	441	359	311	238	173	132	108	81	69	53	46	41	37	34	31
T-200	200	551	449	389	398	216	165	135	102	86	66	57	51	47	42	39
T-240	240	661	538	467	357	260	199	163	122	103	79	68	61	56	51	47
T-280	280	771	628	544	417	303	232	190	142	120	92	80	71	65	59	54
T-300	300	826	673	583	447	325	248	203	152	129	99	85	76	70	64	58
T-320	320	881	718	622	477	346	265	217	163	137	106	91	82	75	68	62
T-360	360	991	808	700	536	389	298	244	183	155	119	102	92	84	76	70
T-400	400	1101	897	778	596	433	331	271	203	172	132	114	102	93	85	78
T-440	440	1211	987	856	655	476	364	298	22	189	145	125	112	102	93	85
T-480	480	1322	1077	933	715	519	397	325	244	206	159	137	122	112	102	93
T-500	500	1377	1122	972	745	541	414	339	254	215	165	142	127	116	106	97
T-520	520	1432	1166	1011	775	563	430	352	264	223	172	148	132	121	110	101
T-560	560	1542	1256	1089	834	606	463	379	284	240	185	159	143	130	119	109
T-600	600	1652	1346	1167	894	649	496	406	305	258	198	171	153	140	127	116

T-650	650	1790	1458	1264	968	703	538	440	330	279	215	185	166	151	138	126
T-680	680	1872	1525	1322	1013	736	563	460	345	292	225	194	173	158	144	132
T-760	760	2092	1705	1478	1132	822	629	515	386	326	251	216	194	177	161	147
T-850	850	2340	1907	1653	1266	920	703	576	432	365	281	242	217	198	180	165
T-925	925	2547	2075	1798	1378	1001	765	626	470	397	305	263	236	215	196	179
T-1000	1000	2753	2243	1944	1489	1082	827	677	508	429	330	285	255	233	212	194

Table 129: VALVE REGULATED LEAD ACID BATTERIES Discharge Characteristics in Watts 1.80 ECV

Module type	Nominal capacity (Ah) at C_{10}	Dischrge in minutes				Dischrge in hours										
		5	10	15	30	1	1.5	2	3	4	5	6	7	8	9	10
T-80	80	216	175	160	116	85	66	53	39	32	26	23	20	18	16	15
T-100	100	270	218	200	145	107	82	66	49	40	32	28	25	22	20	19
T-120	120	323	268	239	174	128	98	79	59	48	39	34	30	27	24	23
T-160	160	431	349	319	232	171	131	105	79	63	52	45	40	36	32	30
T-200	200	539	437	399	290	214	164	132	98	79	65	56	50	45	41	38
T-240	240	647	524	479	348	256	197	158	118	95	78	68	60	54	49	45
T-280	280	755	611	559	406	299	229	185	138	111	91	79	69	63	57	53
T-300	300	809	655	599	435	320	246	198	147	119	97	85	74	67	61	57
T-320	320	862	698	638	464	342	262	211	157	127	104	90	79	72	65	60
T-360	360	970	786	718	522	385	295	237	177	143	117	101	89	80	73	68
T-400	400	1078	873	798	580	427	328	264	197	159	130	113	99	89	81	76
T-440	440	1186	960	878	638	470	360	290	216	174	142	124	109	98	89	83
T-480	480	1294	1048	958	696	513	393	316	236	190	155	135	119	107	97	91
T-500	500	1348	1091	998	725	534	409	330	246	198	162	141	124	112	101	95
T-520	520	1402	1135	1037	754	555	426	343	256	206	168	147	129	116	105	98
T-560	560	1509	1222	1117	812	598	459	369	275	222	181	158	139	125	113	106
T-600	600	1617	1310	1197	870	641	491	396	295	238	194	169	149	134	122	113
T-650	650	1752	1419	1297	942	694	532	429	319	258	210	183	161	145	132	123
T-680	680	1833	1484	1357	986	726	557	448	334	270	220	192	169	152	138	129
T-760	760	2048	1659	1516	1102	812	622	501	374	301	246	214	189	170	154	144
T-850	850	2291	1855	1696	1232	908	696	560	418	337	275	240	211	190	172	161
T-925	925	2493	2019	1846	1341	988	757	610	455	367	300	261	230	207	187	175
T-1000	1000	2695	2183	1995	1450	1068	819	659	492	396	324	282	248	223	203	189

Table 130: VALVE REGULATED LEAD ACID BATTERIES Discharge Characteristics in Amperes To 1.75 ECV

Module type	Nominal capacity (Ah) at C_{10}	Dischrge in minutes				Dischrge in hours										
		5	10	15	30	1	1.5	2	3	4	5	6	7	8	9	10
T-80	80	123.1	98.8	84.2	64	46	34.8	28.4	21.3	16.8	13.8	11.9	10.7	9.6	8.8	8
T-100	100	153.8	123.5	105.3	79	57	43.5	35.5	26.6	21	17.2	14.9	13.3	12	11	10
T-120	120	184.6	148.1	126.3	95	69	52.2	42.6	31.9	25.2	20.7	17.9	16	14.5	13.2	12
T-160	160	246	198	168	127	91	70	57	43	34	28	24	21	19	18	16
T-200	200	308	247	211	159	114	87	71	53	42	34	30	27	24	22	20
T-240	240	369	296	253	190	137	104	85	64	50	41	36	32	29	26	24
T-280	280	431	346	295	222	160	122	99	74	59	48	42	37	34	31	28
T-300	300	462	370	316	238	171	130	106	80	63	52	45	40	36	33	30
T-320	320	492	395	337	254	183	139	113	85	67	55	48	43	39	35	32
T-360	360	554	444	379	286	206	157	128	96	76	62	54	48	43	40	36
T-400	400	615	494	421	317	229	174	142	106	84	69	60	53	48	44	40
T-440	440	677	543	463	349	251	191	156	117	92	76	66	59	53	48	44
T-480	480	738	593	505	381	274	209	170	128	101	83	72	64	58	53	48
T-500	500	769	617	526	397	286	217	177	133	105	86	75	67	60	55	50
T-520	520	800	642	547	413	297	226	184	138	109	90	78	69	63	57	52
T-560	560	862	691	589	444	320	243	199	149	118	96	84	75	67	62	56
T-600	600	923	741	632	476	343	261	213	160	126	103	90	80	72	66	60
T-650	650	1000	802	684	516	371	283	230	173	137	112	97	87	78	71	65
T-680	680	1046	840	716	540	389	296	241	181	143	117	101	91	82	75	68
T-760	760	1169	938	800	603	434	330	270	202	160	131	113	101	92	84	76
T-850	850	1308	1049	895	675	486	370	301	226	179	146	127	113	102	93	85
T-925	925	1423	1142	974	734	529	402	328	246	194	159	138	123	111	102	93
T-1000	1000	1538	1235	1053	794	571	435	355	266	210	172	149	133	120	110	100

Table 131: VALVE REGULATED LEAD ACID BATTERIES
Discharge Characteristics in Amperes To 1.80 ECV

Module type	Nominal capacity (Ah) at C_{10}	Dischrge in minutes				Dischrge in hours										
		5	10	15	30	1	1.5	2	3	4	5	6	7	8	9	10
T-80	80	119.4	95.2	81.6	61.5	44.9	34.2	27.5	20.5	16.5	13.4	11.6	10.3	9.3	8.4	7.
T-100	100	149.3	119	102	76.9	56.2	42.7	34.4	25.6	20.6	16.7	14.5	12.9	11.6	10.5	9.
T-120	120	179.1	142.9	122.4	92.3	67.4	51.3	41.2	30.7	24.7	20	17.4	15.5	13.9	12.6	11.
T-160	160	239	190	163	123	90	68	55	41	33	27	23	21	19	17	15
T-200	200	299	238	204	154	112	85	69	51	41	33	29	26	23	21	19
T-240	240	358	286	245	185	135	103	82	61	49	40	35	31	28	25	23
T-280	280	418	333	286	215	157	120	96	72	58	47	41	36	32	29	27
T-300	300	448	357	306	231	169	128	103	77	62	50	43	39	35	32	29
T-320	320	478	381	327	246	180	137	110	82	66	53	46	41	37	34	31
T-360	360	537	429	367	277	202	154	124	92	74	60	52	46	42	38	35
T-400	400	597	476	408	308	225	171	137	102	82	67	58	52	46	42	39
T-440	440	657	524	449	338	247	188	151	113	91	73	64	57	51	46	42
T-480	480	716	571	490	369	270	205	165	123	99	80	70	62	56	51	46
T-500	500	746	595	510	385	281	214	172	128	103	83	72	65	58	53	48
T-520	520	776	619	531	400	292	222	179	133	107	87	75	67	60	55	50
T-560	560	836	667	571	431	315	239	192	143	115	93	81	72	65	59	54
T-600	600	970	8E+05	612	462	337	256	206	153	124	100	87	77	69	63	58
T-650	650	1015	810	663	500	365	278	223	166	134	109	94	84	75	68	63
T-680	680	1134	905	694	523	382	291	234	174	140	114	99	88	79	72	66
T-760	760	1269	1012	776	585	427	325	261	194	157	127	110	98	88	80	73
T-850	850	1381	1101	867	654	478	363	292	217	175	142	123	110	98	89	82
T-925	925	1493	1190	944	712	520	395	318	237	191	154	134	119	107	97	89
T-1000	1000	1538	1235	1020	769	562	427	344	256	206	167	145	129	116	105	97

Table 132: SIZE OF CABLES, EARTHING CONDUCTORS & PROTECTION FOR STANDBY GENERATORS

Generator capacity kVA	Full load current rating Amps	Cable size AYFY mm^2	Earth conductor size mm^2/SWG	Protection	Panel instruments
5	7	4	8.3/10	MCB	AM, VM & FM
7.5	10.5	4	8.3/10	MCB	
10	14	4	8.3/10	MCB	
12.5	17.5	6	8.3/10	MCB	AM, VM, FM & EM
15	21	10	8.3/10	MCB	
20	28	10	8.3/10	MCB	
25	35	16	18.6/6	MCB	
30	42	16	18.6/6	MCB	
35	49	25	18.6/6	MCB	
40	56	35	27.27/4	MCB	
45	63	35	27.27/4	MCB	
50	70	35	27.27/4	MCCB	
63	88	50	27.27/4	MCCB	
82.5	115	95	25 × 3	MCCB	
100	140	120	25 × 3	MCCB + Stand by low Set earth fault Relay using Ct in neutral Earthing conductor	AM, VM, FM & EM
125	175	150	25 × 3		
160	224	185	25 × 3		
180	252	300	25 × 3		
200	280	2 × 120	25 × 3		
225	315	2 × 225	25 × 3		
250	350	2 × 185	25 × 3		
320	448	2 × 300	25 × 3		
400	560	2 × 400	25 × 3		
500	700	3 × 400	25 × 3	ACB with Overload and E/F release	AM, VM, FM EM, PFM, kVAM
625	875	4 × 400	25 × 3		
750	1050	Busbar Trunking	25 × 3		

For 1000 kVA and above, ACB with thermal O / L, voltage controlled O / C relay, over voltage, under voltage, negative sequence, low set stand by earth fault relays and REF/ Different relays with fuel shutoff facilities. Over speed protection shall be provided for the engine.

(Source: KSEIE Handbook)

Table 133: PRIME POWER RATING OF GENERATING SETS (40-125 kVA)
(Source: Kirloskar Oil Engines Ltd)

Prime rating at rated rpm (as per ISO 8528)[1]		kVA	**40**	**40**	**45**	**62.5**	**62.5**	**82.5**	**100**	**125**	**125HD**
		kW	32	32	36	50	50	60	80	100	100
Genset Model		–	KG1-40AS	KG1-40WS	KG1-45WS	KG1-62.5AS	KG1-62.5AS	KG1-82.5WS	KG1-100WS	KG1-125WS	KG1-125WS HD
Frequency		Hz	50	50	50	50	50	50	50	50	50
Power factor		lagging	0.8	0.8	0.8	0.8	0.8	0.8	0.8	0.8	0.8
Voltage		V	230 (1Ø) & 415 (3Ø)					415(3Ø)			
Governing class (As per ISO 8528 Part-V)		–	G2	G2	G2	G2	G2	G2	G3	G3	G3
Noise level		dBA	< 75	< 75	< 75	< 75	< 75	< 75	< 75	< 75	< 75
Fuel Consumption*	At 100% Load	Ltrs/hr	9.9	9.2	10.3	16.2	14.1	18.8	21.9	27.4	27.5
	At 75% Load		7.9	7.4	8.7	12.7	11.3	13.8	16.9	20.2	21.4
	At 50% Load		5.8	5.5	5.9	8.9	7.5	9.9	12.2	15.3	15.3
Fuel tank capacity		Ltrs	100	100	100	150	150	230	230	230	50
Weight of genset with canopy (approx..)^	Dry	Kg	1270	1250	1250	1700	1420	1710	2040	2090	2520
	Wet	Kg	1370	1335	1335	1825	1550	1840	2240	2290	2730
Overall dimensions of genset^	Length	mm	2400	2550	2550	3150	2800	3200	3220	3220	3650
	Width	mm	1050	1050	1050	1200	1100	1100	1300	1300	1350
	Height	mm	1670	1450	1450	1635	1595	1595	1795	1795	1915
Electrical battery starting voltage		Volts-DC	12	12	12	12	12	12	12	12	12

ENGINE										
Engine Model	–	HA494TC IG1	3R1040T AG1	3R1040T AG1	HA694TC IG1	4R810T AG1	4R810T AG1	4K1080T AG2	4K1080T AG2	6R1080T AG1
Rated output (Prime Continuous rating as per ISO 8528-1)	kW	41.19	41.2	41.2	61	61	74.8	114.7	114.7	114.7
	HP	56	56	56	83	83	102	156	156	175
No. of cylinder	Number	4	3	3	6	4	4	4	4	6
Cubic capacity	Ltrs	3.77	3.12	3.12	5.65	3.24	4.16	4.32	4.32	6.49
Bore × Stroke	mm	100 × 120	105 × 120	105 × 120	105 × 120	100 × 120	96 × 112	105 × 120	105 × 125	105 × 125
Rated Speed	RPM	1500	1500	1500	1500	1500	1500	1500	1500	1500
Aspiration	NA/TC/TA	TC	TA	TA	TC	TA	TA	TA	TA	TC
Lube Oil change period	Hrs	500	500	500	500	500	500	500	500	500
Lube Oil Sump Capacity	Ltrs	8.3	8	8	11	10	10	14	14	18
Coolant Capacity	Ltrs		11.5	11.5		117.5	24	21	21	28
ALTERNATOR										
Insulation Class	–	Class H								
Alternator Efficiency (at 100% load) 0.8 pf**	%	89.2	89.2	89.2	91	91	91.4	92	92.5	92.5
Max. Voltage Dip at Full Load 0.8 pf Lag	Sec	< 20%	< 16%	< 16%	< 20%	< 20%	< 20%	< 20%	< 20%	< 20%
Max. Time to build up rated voltage at Rated RPM		< 5 sec provided engine reaches the rated speed					< Voltage recovery time 6 sec			

Table 134: IMPORTANT INDIAN STANDARD SPPECIFICATIONS

ISS number	Year	Title
325	1978	Three-phase induction motors
335	1993	New insulating oils
375	1963	Marking arrangements for switchgears, busbars, main connections and auxiliary wiring. (Supersded by IS: 5578 and IS: 11353)
694	1990	PVC insulated cables for working voltage up to and including 1100 V.
722	1986	AC Electricity meters
732	1989	Code of practice for wiring installations
900	1992	Code of practice for installation and maintenance of induction motors
1180 (1-2)	1989	Outdoor type three-phase distribution transformers up to and including 100 kVA
1248 (1-9)	1993	Direct acting analog electrical measuring instruments Parts 1 to 9
1255	1983	Installation and maintenance of power cables up to and including 33 kV
1336	1959	Recommendation of the colour of push buttons (superseded by IS:6875-2)
1554(1-2)	1988	PVC insulated (Heavy duty) electric cables – Parts - 1, 2
1860	1980	Code of practice for installation, operation and maintenance of passenger and goods lifts
1866	1983	Code of practice for maintenance and supervision of mineral insulating oil in equipment
1876	1961	Method of voltage measurements by means of sphere gaps.
1944 (5-7)	1981	Code of practice for lighting of public thoroughfares
2071 (1-3)	1993	Methods of high voltage testing
2026 (1-4)	1981	Power transformers
2309	1989	Code of practice for the protection of buildings and other allied structures against lightning
2675	1983	Enclosed distribution boards and cut outs for voltages not exceeding 1000 V.
2705 (1-3)	1992	Current transformers
2992	1987	Hand operated insulation resistance testers
2993	1975	Motor capacitors
3043	1987	Code of practice for earthing
3151	1982	Earthing transformers
3231	1965	Specification for electrical relays for power system protection

3528	1966	Water proof electric light fittings.
3553	1966	Water tight electric light fittings
3961 (1-5)	1967	Recommended current ratings of cables-Parts 1 to 5
4029	1967	Guide for testing of three-phase induction motors
4146	1983	Application guide for voltage transformers
4288	1988	PVC insulated (heavy duty) electric cables with solid aluminium conductors
4347	1967	Code of practice for hospital lighting
4591	1968	Installation and maintenance of escalators
4648	1968	Guide for electrical layout in residential buildings
4665 (1, 2)	1984	Portable electric motor operated tools
4728	1975	Terminal markings and direction of rotation for electrical machinery
4889	1968	Method of determining efficiency of rotating electrical machines
4894	1987	Centrifugal fans
5216 (1, 2)	1982	Guide for safety procedures in electrical work
5579	1985	Neon testers
5834 (1-3)	1994	Electrical timer relays for industrial purpose
6103	1971	Method of test for specific resistance of electrical insulating liquids
6104	1971	Test for interfacial tension oil against water by the ring method
6209	1982	Methods of partial discharge measurements
6262	1971	Designation of methods of cooling of electrical machines
6272	1987	Industrial cooling fans
6381	1972	Construction and testing of electrical apparatus with type 'e' protection
6665	1972	Code of practice for industrial lighting
6792	1992	Method of determination of electric strength of insulating oils
6855	1993	Method of sampling of liquid dielectrics
6875	1973	Control switches for voltages up to 1000 V AC and 1200 V DC
7098 (1-2)	1988	XLPE insulated PVC sheathed cables
7132	1973	Guide for testing synchronous machines
7372	1995	Lead acid storage batteries for motor vehicles
7537	1974	Road traffic signals
7816	1975	Guide for testing of insulation resistance of rotating machines
8061	1976	Code of practice for design, installation and maintenance of service lines up to and including 650 V

8084	1976	Interconnecting busbars for AC voltages above 1 kV and up to and including 36 kV.
8289	1976	Electrical equipment with type 'n' protection
8320	1982	General requirements and method of test for lead acid storage batteries
8437 (1-2)	1993	Guide on effects of current passing through the human body
8468	1977	On load tap changers
8478	1977	Application guide for on load tap changers
8623 (1-3)	1993	Specification for LV switchgear assemblies
8686	1977	Static protective relays
8789	1986	Values of performance characteristics for three-phase induction motors
8828	1995	Circuit beakers for over current protection for household and similar installations
8923	1978	Warning symbols for dangerous voltages
8935	1985	Electric solenoid operated actuators
9124	1979	Guide for maintenance and field tests for electrical relays
9223	1989	Portable earth resistance meters
9283	1979	Motors for submersible pump sets
9434	1992	Guide for sampling and analysis of free and dissolved gases and oil from oil filled equipment
9677	1980	Limits of temperature rise of winding of electrical equipment when tested by different methods
9678	1980	Methods of measuring temperature rise of electrical equipment
9792	1987	Guide for testing, calibration and maintenance of AC electricity meters
10027	1981	Composite units of air break switches and re-wirable fuses for voltages not exceeding 650 V
10028 (1-3)	1985	Code of practice for selection, installation and maintenance of transformers
10118 (1-3)	1982	Code of practice for selection, installation and maintenance of switchgears
10561	1983	Application guide for power transformers
10810	1984	Method of testing of cables
10894	1984	Code of practice for lighting of educational institutions
11171	1985	Dry type power transformers
11260 (1-2)	1985	Stabilized power supplies AC output
11599	1986	Portable AC clip-on ammeters

11633	1986	Specification for lift doors
11994	1986	Portable insulation resistance meters
12021	1987	Current transformers for switchgear and control gear, voltage not exceeding 1000 V AC
12065	1987	Permissible limits of noise level of rotating electrical machines
12083 (1-2)	1983	Electrical relays
12188	1987	Electric direct arc melting furnaces
12615	1989	Energy efficient three-phase induction motors
12640	1988	Residual current operated circuit breakers
12802	1989	Temperature rise measurement of rotating electrical machines
12824	1989	Type of duty and classes of rating assigned to rotating electrical machines
13010	1990	AC watt-hour meters class 0.5 1.0 and 2
13032	1991	Miniature circuit breaker boards for voltages upto and including 1000 V AC
13067	1991	Impregnation of power capacitors
13107	1991	Guide for measurement of winding resistance of an AC machine during operation at alternating voltage
13118	1991	General requirements of circuit breakers for voltages above 1000 V
13234	1992	Guide for short circuit calculation in three-phase AC systems (IEC 60909)
13235	1991	Calculation of the effects of short circuit currents (similar to IEC 60865)
13364 (1-2)	1992	Specification for AC generators driven by reciprocating IC engines
13529	1992	Effects of unbalanced voltage on the performance of three-phase induction motors
13555	1993	Guide for selection and application of three-phase induction motors for different types of driven equipment
13703 (1-2)	1993	LV fuses for voltage not exceeding 1000 V AC and 1500 V DC
13779	1990	AC watt-hour meters
13947 (1-5)	1993	LV switch gear and control gear, Parts 1-5
14372	1996	Volt ampere hour meters for full power factor range
14415	1997	Volt ampere hour meters for restricted power factor range (superseding IS 722-5)
14665 (1-5)	2000	Electric traction lifts

Table 135: NATIONAL/INTERNATIONAL ORGANISATIONS FOR STANDARDISATION

Abbreviation	Description	Website
IEC	International Electrotechnical Commission	www.iec.ch
ISO	International Standards Organization	www.iso.org
CEN	European Committee for Standardization (Comite Europeen de Normalisation)	www.cen.eu
CENELEC	European Committee for Electrotechnical Standardization (Comite Europeen de Normalisation Electrotechnique)	www.cenelec.eu
IEEE	Institution of Electrical & Electronics Engineers	www.ieee.org
ANSI	American National Standards Institution	www.ansi.org
AS	Australian Standards Organisation	www.standards.org.au
BS	British Standards Organisation	www.bsigroup.com
JSA	Japanese Standards Organisation	www.jsa.or.jp
NEN	Netherland Standards Organisation	www.nen.nl
BIS	Bureau of Indian Standards	www.bis.org
CSA	Canadian Standards Association	www.csa.ca
CEI	Italian Electrotechnical Committee (Comitato Electrotechnico Italiano)	www.ceiuni.it
NBN	Belgian Standards Institution	www.nbn.be
NEN	Netherlands Standards Institution (Nederlands Normalisatie Instituut)	www.nen.nl
SEN	Swedish Electrotechnical Standards (Svenska Elektrotekniska Normer)	www.elstandard.se
SEV	Swiss Electrotechnical Association	www.electrosuisse.ch
UTE	French Electrotechnical Association (Union Technique de I'Electricite)	www.afnor.org